D0897795

Energy Efficiency

Building a Clean, Secure Economy

The Hoover Institution gratefully acknowledges
THOMAS AND BARBARA STEPHENSON
for their significant support of the

**Shultz-Stephenson Task Force
on Energy Policy**

and this publication.

ENERGY EFFICIENCY

Building a Clean, Secure Economy

James L. Sweeney

HOOVER INSTITUTION

SHULTZ-STEPHENSON TASK FORCE ON
Energy Policy

STANFORD UNIVERSITY

HOOVER INSTITUTION PRESS

Stanford University Stanford, California

www.hoover.org

Hoover Institution Press Publication No. 668

Hoover Institution at Leland Stanford Junior University,
Stanford, California 94305-6003

First printing 2016
22 21 20 19 18 17 16   8 7 6 5 4 3 2

Manufactured in the United States of America

The paper used in this publication meets the minimum Requirements of the American National Standard for Information Sciences—Permanence of Paper for Printed Library Materials, ANSI/ NISO Z39.48-1992. ∞

Library of Congress Cataloging-in-Publication Data

Names: Sweeney, James L., author.
Title: Energy efficiency : building a clean, secure economy / James Sweeney.
Other titles: Hoover Institution Press publication ; 668.
Description: Stanford, CA : Hoover Institution Press, 2016. | Series: Hoover Institution Press publication ; no. 668
Identifiers: LCCN 2016021959 | ISBN 9780817919542 (clothbound : alk. paper) | ISBN 9780817919566 (epub) | ISBN 9780817919573 (mobipocket) | ISBN 9780817919580 (pdf)
Subjects: LCSH: Energy consumption—United States. | Energy consumption—Economic aspects—United States. | Energy policy—United States.
Classification: LCC HD9502.U52 S94 2016 | DDC 333.79/130973—dc23
LC record available at https://lccn.loc.gov/2016021959

Contents

List of Figures and Tables

Figures

Foreword

The world, let alone the United States, is in the midst of a revolution in the energy area. Previously low-income countries are experiencing economic development and are in need of energy to power that development. There is increasing awareness of the importance of the atmospheric effects of the use of energy. This is not only a matter of climate; simply ask China about the importance of the air we breathe. Then, of course, energy is tightly linked to our security.

Jim Sweeney has been working on this subject for many years, in particular on what can be done to use energy more efficiently. This book presents solid information on this issue and the result is stunning. The contribution of energy efficiency to date in achieving our goals has been dramatic.

What is the cleanest energy around? The energy that is not used. What is the least expensive energy around? The energy that is not used. What is the most secure energy around? The energy that is not used. So energy efficiency is a triple play.

There are many ways in which the efficient use of energy takes place, including simply as a matter of awareness. I remember a time when we had a crisis in electricity in California. In the building where I worked, we slightly dimmed the lights in the corridors, which moderated their glare and actually made the light more comfortable. We were instructed to turn off lights in offices that were not being used. The result was a 13 percent savings in electricity, even though nothing had been invented to alleviate the problem and nobody had been inconvenienced in any way.

But certainly one of the big motivators of efficiency in the use of energy is the price. One of the really interesting aspects of the charts that Jim Sweeney presents in this book is the inflection point almost across the board in the year 1973.

I remember that year well. I was Secretary of the Treasury, having earlier warned that the main threat of the uneasy situation in the Middle East was that of energy supply. Along comes the Arab Oil Boycott in retribution for action by the United States to resupply Israel at the time of the Yom Kippur War. The result made a deep impression throughout the American body politic. Christmas lights were discouraged and gas stations were closed on weekends, so the reality of energy's strategic importance was dramatic in a unique and powerful way.

While prices came down somewhat as the boycott ended, they rose again in 1979 at the time of the Iranian revolution, and in the last decade or so, until very recently, they have been extraordinarily high.

The 1973 inflection point is a clear and dramatic piece of evidence of the ability of people to react to high prices and short supplies. But this book provides much more evidence of reactions and measures taken. Widespread use of LED lighting, for example, will have a major impact on the amount of electricity used.

This book presents a sharp picture showing that the contribution of energy efficiency to the US energy scene has been more powerful than any other single development, in fact more powerful than *all* increases of domestic energy supply taken together. I say this not to downplay the importance of alternative sources of energy but simply to urge continued attention to the importance of using this energy as efficiently as possible.

In my own case, for example, I have solar panels on the roof of my energy-efficient house at Stanford and I drive a very energy-efficient electric car. I have long since paid for my panels by savings on my electricity bill. The electricity used by my car is far less than what is produced by the solar panels, so I am, in effect, driving on sunshine. What's not to like?

I have long advocated a revenue-neutral carbon tax, and the picture presented in this book is evidence that people respond to prices. So let's keep our eye on what has worked and what will continue to work. This book by Jim Sweeney is full of ideas and evidence that can help make the future better than the past.

George P. Shultz
The Thomas W. and Susan B. Ford Distinguished Fellow
Hoover Institution, Stanford University
Stanford, California

Acknowledgments

The Precourt Energy Efficiency Center (PEEC) was founded in October 2006 at Stanford University by a generous gift from Stanford alumnus Jay Precourt. As a Stanford University research institute, PEEC draws upon intellectual resources from the entire university in order to improve opportunities for and implementation of energy efficient technologies, systems, and practices, with an emphasis on economically attractive deployment. Financial, moral, organizational, and intellectual support by Jay Precourt continues to make PEEC viable. PEEC is part of another Stanford organization, also funded initially by Jay Precourt, the Precourt Institute for Energy.

If the Precourt Energy Efficiency Center didn't exist, this book would not, and probably could not, have been written.

The Hoover Institution's Shultz-Stephenson Task Force on Energy Policy addresses energy policy in the United States and its effects on our domestic and international political priorities, particularly our national security. It was Secretary George Shultz, the leader of this task force, who strongly encouraged me to write a book and who frequently kept reminding me to continue the progress. George Shultz has been my inspiration for many decades.

Funding from the Exxon Mobil Corporation, through the Stanford Institute for Economic Policy Research, provided partial financial support.

Finally, thanks to the Energy Information Administration (EIA) of the US Department of Energy. The rich array of quality data, available on the EIA website, has been essential to quantifications in this book.

Many people provided invaluable assistance as this book was under development. My wife, Susan Sweeney, has kept encouraging me, commented on previous drafts, and put up with me during the writing process. Many other people have provided helpful comments, ideas, critiques, intellectual assistance, and language for the previous drafts. Thank you Christina Angelides, Carrie Armel, Jeff Bingaman, Ralph Cavanagh, Ross Chanin, John Conomos, Danny Cullenward, David Fedor, Wendy Fok, Mark Golden, David Goldstein, Kit Kennedy, Jonathan Koomey, Scott Litzelman, Tony Malkin, Robert Marks, Jay Precourt, Gabriel Rosenthal, Ansuman Sahoo, Andreas Schäfer, George Shultz, Michael Sivak, Daniel Sperling, Jerry Sweeney, Margaret Taylor, Roland Wang, John Weyant. And I would especially like to thank George Sweeney and Dian Grueneich for their extensive very valuable comments on previous drafts.

Introduction

I n 1973, the energy world was fundamentally altered by the oil embargo and tripling of world oil prices, with high oil prices remaining for over ten years. The United States turned attention to reducing oil imports, driven by national security and economic concerns. More recently, with recognition of global climate change and carbon dioxide emissions from fossil fuel use, attention has turned toward reducing greenhouse gas emissions, driven by environmental concerns. Energy policy discussions since 1973 have thus centered on energy impacts to three complex and crucial systems: the economy, the environment, and security, the "energy policy triangle."

Energy efficiency—defined here as economically efficient reductions in energy use—reduces energy intensity of the economy—defined as the energy consumption per constant dollar of gross domestic product (GDP). This in turn benefits national security, the environment, and the economy. This energy policy context is the subject of Chapter 1 of this book.

Since the oil embargo, individuals, corporations, and other organizations have found ways of economically reducing energy use so that energy efficiency is now all around us, as discussed in Chapter 2. For example, a high-quality LED lightbulb uses only 11 watts to provide the same amount of light as a 60-watt incandescent lamp. The consumer recovers the higher cost of the LED bulb many times over while helping the environment and energy security. A refrigerator purchased today uses less than a third of the energy used by a 1973 refrigerator. New cars get about twice the mileage of cars on the road in 1973. Air travel uses about one-quarter as much fuel per passenger mile as in 1970. Companies have adopted data-driven methods for finding energy inefficiency and for reducing energy use. Energy-efficiency upgrades for office buildings have been coupled with landlord/tenant collaborations on energy and new contractual structures. Companies use behavioral strategies to motivate energy-efficiency practices by their employees. Many corporations have adopted internal carbon pricing systems to motivate internal changes.

These are but examples of energy-efficiency changes all around us. US economy-wide data allow a quantification, in Chapter 3, of energy efficiency impacts on overall energy use. An analytical challenge is that although we readily measure how much energy is produced, less clear is the measurement of energy *not* used because of energy-efficient technologies and practices. Thus, to quantify impacts of enhanced energy efficiency, I first estimate what energy use would have been with less energy-efficiency gains, by constructing a "limited-energy-efficiency" benchmark based on the modest rate of efficiency gains occurring before the crisis. Pre-crisis, little attention was paid to energy efficiency; energy use was growing slightly less than economic growth.

Faced with the high cost of imported oil and the threat of further oil embargos, federal and state governments adopted policies to increase domestic supply of energy and to reduce energy intensity of the economy. In the private sector, possible reductions in energy use became a factor in products and processes. Individual households starting changing their choices.

These factors together started and sustained a slow cumulative process of energy-efficiency improvements. Economy-wide energy intensity began decreasing substantially faster than historically. From 1973 until the oil price collapse in 1985, with high energy prices, many energy policy initiatives, and high recognition of energy use, energy intensity declined on average 2.7 percent per year. Once the world oil price collapsed, public recognition of energy security moved toward the background. But concerns about global climate change grew. Energy policy activity slackened but did not disappear. From 1985 until now, on average energy intensity has declined 1.7 percent per year.

Annual changes of 2.7 percent or 1.7 percent per year accumulated over four decades. The net result: the US economy-wide energy intensity decreased by 57 percent, from 14,000 BTUs per dollar of GDP in 1973 to 6000 BTUs per dollar of GDP in 2014 (both figures in 2009 dollars). Some reduction would have occurred under the limited-energy-efficiency benchmark. But most reductions resulted from enhanced energy efficiency. Energy efficiency has limited energy use to 100 quadrillion BTUs per year rather than the 180 quadrillion BTUs per year projected absent enhanced energy efficiency.[1]

Those large results often go unrecognized, partially because energy efficiency progress has been an accumulation of small changes, broadly distributed, often invisible to outsiders.

These changes in energy use have been fundamental to US energy security and US greenhouse gas emissions, discussed in Chapter 4. Since the energy crisis of 1973–74, US energy efficiency has done more to curb greenhouse gas emissions and more to reduce net energy

---

1. The observation that there has been a dramatic role of energy efficiency is absolutely not original to this book, but has been made by many other authors. Art Rosenfeld, "grandfather of energy efficiency," is probably the best-known and first person to promote the concept of energy efficiency. One of Art's mentees, Amory Lovins, for decades has been praising energy efficiency as "generally the largest, least expensive, most benign, most quickly deployable, least visible, least understood, and most neglected way to provide energy services." A.B. Lovins, "Energy Efficiency, Taxonomic Overview," *Encyclopedia of Energy*, 2:383–401 (2004), San Diego, CA, and Oxford, UK: Elsevier. More recently, the conclusion is well articulated in the National Research Council report *America's Energy Future* and by the American Council for an Energy-Efficient Economy (ACEEE): "Energy Efficiency in the United States: 35 Years and Counting," June 30, 2015, Steven Nadel, Neal Elliott, and Therese Langer, Research Report E1502. The US Department of Energy publishes data on energy indicators at http://www1.eere.energy.gov/analysis/eii_index .html. Marilyn Brown and Yu Wang recently published *Green Savings: How Policies and Markets Drive Energy Efficiency* in this area. Many other documents have communicated the conclusion that energy efficiency has been fundamentally important to the US energy system.

imports than have increases in domestic production of oil, gas, coal, geothermal energy, nuclear power, solar power, wind power, and biofuels—all put together.

Impacts of energy efficiency and of domestic energy production changes are quantified in Figure Intro.1, which plots US energy consumption, domestic supply, and net imports of primary[2] energy from 1950 through 2014. The difference between total energy consumed (shown as a blue line) and total domestic supply of energy (the black, purple, and green areas) gives net energy imports. Actual net energy imports are shown as a light gray area, and the net energy imports under the limited-energy-efficiency benchmark are shown as the light gray area plus the dark gray area, given the actual levels of domestic primary energy production.

Net energy imports have declined to just below the 1973 level. Domestic energy production from all US sources together has increased by 24 quadrillion BTUs per year, as shown by the gold arrow; increases in efficiency have reduced consumption of energy from what it would have been otherwise by 80 quadrillion BTUs per year—well over three times the increase in all forms of domestic energy production. Energy efficiency has put the United States on the road to net energy sufficiency, greatly enhancing security.

Net energy imports into the United States would not have actually increased to 100 quadrillion BTUs absent the enhanced energy efficiency. World energy markets would not have sustained such an increase. Such high imports would have been politically unacceptable. The high level would have triggered large energy price increases for imported

---

2. Data in this report are of primary energy consumption or production, all measured in terms of quadrillion BTUs of energy (10^{15} BTUs). Primary energy includes crude oil, natural gas, coal, geothermal energy, nuclear power, solar energy, wind energy, and organic material for biofuels. Primary energy sources are converted to "secondary energy," or "energy carriers" such as electricity or refined fuels.

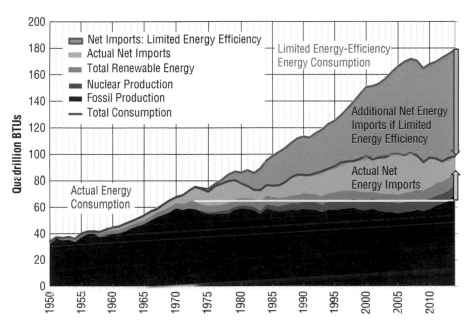

Figure Intro.1. US Net Energy Imports: Actual vs. Limited Energy Efficiency

energy, increases in domestic production of all primary energy sources, increases in deployment of electric generators, and energy-efficiency initiatives. Even with these changes, absent enhanced energy efficiency, there would have been large increases in energy imports, particularly petroleum and possibly coal, greatly reducing energy security. The combination of some increased energy production and much energy efficiency has avoided those serious consequences.

Energy efficiency has also been dominant in reducing carbon dioxide emissions. The carbon intensity of the economy—carbon dioxide emissions per constant dollar of GDP—can be decomposed into two multiplicative factors: the carbon intensity of energy consumption and the energy intensity of the economy. The carbon intensity of energy consumption is determined by the fractions of energy consumed from various energy sources (e.g., oil, gas, coal, hydropower, nuclear, wind, solar, geothermal). The greater the fraction of low-carbon or zero-carbon energy, the lower the carbon intensity of energy consumption.

Normalized relative to 1973 levels, so the carbon intensity of the economy is shown as 1.0 in 1973, US data are graphed from 1950 through 2014 in Figure Intro.2. This graph shows the percentage change in carbon intensity of the economy from 1973 decomposed into the three components: the pre-1973 energy-intensity trend, shown as a blue area; impacts of enhanced energy efficiency since 1973, shown as a green area; and impacts of decarbonization of energy consumption, shown as a gray area.

These three factors taken together have reduced carbon intensity of the US economy to 39 percent of its 1973 intensity, as shown by the red line labeled "Carbon Intensity of the Economy."

Since 1973, US energy-intensity reductions (the green plus the blue areas in Figure Intro.2) have been about nine times as important as reductions in carbon intensity of energy consumption, for reducing the carbon intensity of the economy. Enhanced energy-efficiency changes have been about six times as important as reductions in carbon intensity of energy consumption.

Chapter 5 provides more detail, discussing energy-use changes in the four major energy consuming sectors—residential, transportation, commercial, and industrial—and shows that enhanced energy efficiency was not concentrated in any single part of the economy, but rather was broadly distributed throughout all major energy-consuming sectors. Special attention is given to the industrial sector, where about half of the energy-intensity reductions were due to structural shifts in the economy and about half to reductions within existing industries.

What were the forces behind these remarkable changes? Chapter 6 shows that the cumulative, broadly distributed growth in energy effi-

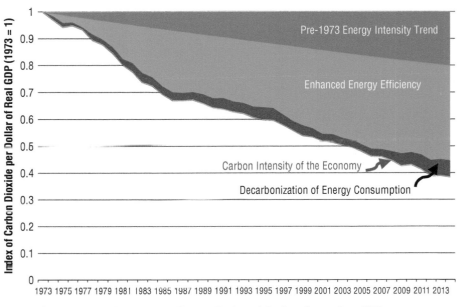

Figure Intro.2. Factors Leading to Reduced Carbon Intensity of US Economy

ciency resulted from many forces working together. Energy prices increased throughout the economy. Attitudes about energy changed. Awareness of energy problems—insecurity of energy imports and environmental degradation from greenhouse gas emissions— heightened. Energy efficiency became a profit strategy. Private-sector and public-sector entities innovated in technology and management, often made possible directly or indirectly by R&D. The federal government and several state governments established energy-efficiency regulations and implemented information programs, labeling, and other nudges. Many utilities launched programs to encourage energy-use reduction. Governmental and utility-based subsidies sped up technology diffusion. Existing non-governmental organizations turned their attention to energy efficiency and new such organizations were created.

Efforts to attribute changes to just one factor—be it market competition, regulation, higher prices, increased awareness, utility programs, or nudges—miss the important lesson from energy-efficiency history. These factors were mutually reinforcing in energy-efficiency gains. And these gains did not occur in simply one place. Rather they were broadly distributed in companies, government agencies, households, and in the transportation system.

The policy lessons from the last forty years give some guidance for moving forward, the subject of Chapter 7. Looking ahead, this same basic configuration of forces is important for continuing benefits of energy efficiency. Although many barriers still exist to full economically efficient reductions in energy use, market failures, structural impediments, and behavioral issues can be addressed by creating instruments matched to the various barriers.

The trends of decreasing energy intensity can readily be preserved. With careful nurturing of private- and public-sector energy efficiency, with public awareness, with appropriate pricing, with appropriate policies, and with increased R&D, the trend of increasing energy efficiency can be accelerated, with further beneficial impacts on the environment, national security, and the United States economy.

1

The Policy Context for Energy Efficiency

President Obama, in his 2013 State of the Union address, reviewed US energy progress since he took office and described an "all-of-the-above" approach for further progress. The accompanying White House Fact Sheet focused on domestic energy production and its impacts on the economy, energy security, and the environment:

> Since President Obama took office, oil and gas production has increased each year, while oil imports have fallen to a 20-year low; renewable electricity generation from wind, solar, and geothermal sources has doubled; and our emissions of the dangerous carbon pollution that threatens our planet have fallen to their lowest level in nearly two decades. In short, the President's approach is working. It's a winning strategy for the economy, energy security, and the environment.[1]

Although the president's summary statements focused on the domestic supply of energy, the administration did give some recognition to energy efficiency in cars and trucks. Nevertheless, Obama appeared to ignore other energy-efficiency improvements that had been continuing for decades:

> The Environmental Protection Agency (EPA) has released a new report that underscores the progress we have made to improve fuel

---

1. FACT SHEET: President Obama's Blueprint for a Clean and Secure Energy Future, press release, The White House, Office of the Press Secretary, March 15, 2013, https://www
.whitehouse.gov/the-press-office/2013/03/15/fact-sheet-president-obama-s-blueprint-clean
-and-secure-energy-future.

economy, save American families money at the pump, and reduce carbon pollution that contributes to climate change.[2]

The White House Fact Sheet, like the vast majority of energy statements from members of both political parties, paid scant attention to the many energy-efficiency enhancements since the energy crisis of 1973–74. The supply side of energy markets again was the focus of political discourse, even though most of the action has been on the demand side of these markets.

In fact, these historical changes in energy efficiency have had more beneficial impacts on US energy security and on the environment than all of the increases in domestic production of oil, gas, coal, geothermal energy, nuclear power, solar power, wind power, and biofuels combined.

This book reviews the advances in energy efficiency within the United States since the 1973–74 energy crisis and documents the generally unheralded contributions of energy efficiency to the economy, energy security, and the environment. This history suggests that the United States can enjoy the many benefits of future efficiency gains if it maintains the interacting conditions that enabled accomplishments of the past 40 years. With careful nurturing of private- and public-sector energy efficiency, with appropriate pricing and other policies, the trend of increasing energy efficiency can be accelerated, with further beneficial impacts on the environment, national security, and the economy.

Energy Efficiency as an Energy Policy Strategy

To understand why energy efficiency—economically efficient reductions in energy use—is a desirable policy strategy, one needs merely to reflect on the basic goals underlying United States energy policy, at

---

2. Ibid.

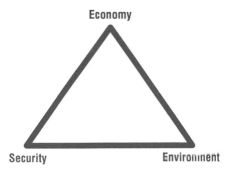

Figure 1.1. The Energy Policy Triangle

least since the energy crisis of 1973–74.[3] These goals have been adopted not just by the US, but internationally, although different countries place different emphases on the three goals.

Energy policy typically strives for improvements in the health and growth of the economy, protection of the domestic and international environment, and enhancement of domestic and international security. These three policy goals are summarized by the energy policy triangle (Figure 1.1).

The Environment

Energy-related issues of the environment include local and international impacts of energy production and use. Energy efficiency is good for the environment: reductions in energy use mean that less energy needs to be produced, transported, or transmitted. Because virtually all energy production and transportation have negative environmental consequences, energy efficiency reduces environmental impacts. Stated most succinctly: the cleanest energy is the energy you don't need in the first place.

---

3. As seen from the White House Fact Sheet, these have been adopted by President Obama.

One particularly important environmental issue is the role of energy efficiency in reducing greenhouse gas emissions in the atmosphere, either carbon dioxide or methane. Combustion of fossil fuels is the dominant source of greenhouse gas emissions in the United States. In a later chapter of this book, I quantify the role of changing intensity of energy use and that of energy system decarbonization.

Other domestic environmental impacts include release of particulates, oxides of nitrogen, sulfur dioxide, and other pollutants into the air[4]; water pollution associated with the production of fossil fuels such as coal or natural gas; use of water for production of primary energy or for conversion of primary energy into electricity; and disruption of natural habitats.

Security

Issues of security include vulnerability of the United States to deliberate or accidental restrictions on oil imports, vulnerability of the US economy to rapid fluctuations in energy prices, or limitations placed on US foreign-policy options as a result of US dependence on foreign energy sources. Domestic security issues include protection against terrorist attacks on energy infrastructure such as the electric grid, or impacts of natural disruptions such as hurricanes or other storms.

Energy efficiency is beneficial for security. For at least 60 years, the United States has been a net importer of energy, particularly oil. Subsequent to the energy crisis, energy efficiency has allowed the United States to reduce energy imports so that the United States may soon become self-sufficient in energy, with energy efficiency having a larger role than domestic energy supply increases. Reductions in electricity use

---

4. Typically, energy efficiency reduces these emissions, but some changes, for example, diesel vehicles, can make some impacts worse while making others better.

can reduce stress on transmission lines and congestion in the grid and can increase flexibility of the existing grid by reducing the time variability of electricity use, and therefore can increase domestic security.

In this book, data from the US Energy Information Administration is used to decompose changes in the net energy imports, separating changes associated with increases in energy efficiency from those associated with increases in the domestic production of energy.

The Economy

Economic issues include growth of gross domestic product (GDP), the number and quality of jobs available for the population, and the distribution of wealth.

Economically efficient reductions in energy use are, by definition, good for the aggregate economy: economic benefits exceed the economic costs. Reductions in the government use of energy can provide savings for the federal government, reduce the federal deficit, and reduce the balance of trade deficit. Cost-effective reductions in the use of energy by businesses can make them more profitable and thus increase GDP. And cost-effective reductions in the use of energy by households can leave more disposable income available for other purposes. Particularly for low-income people, reductions in energy costs can be important for overall well-being. Reduced use of electricity has avoided the need to construct thousands of megawatts of new electric generation stations, saving billions of dollars.

But not all energy-consumption reductions are necessarily cost-effective. Growth in energy use is typically associated with economic development. Limiting the development of third-world countries so as to reduce energy growth would not be cost-effective. These observations suggest the need to be clear about the definition of "energy efficiency."

Some Terminology: Energy Efficiency, Energy Conservation, Energy Intensity, Energy Productivity

Four terms are used here to describe changes in the energy used in the economy: energy efficiency, energy conservation, energy intensity, and energy productivity. The first two—energy efficiency and energy conservation—typically describe *changes* or *differences* in the way energy is used or the amount of energy used. These are often applied to a device, a process, a business activity, or a personal choice. The second two—energy intensity and energy productivity—provide *aggregate measures* of the total energy used, relative to the size of the economy.

For example, replacing a 60-watt incandescent lightbulb with an 11-watt LED light[5] that gives the same quality of light *changes* the amount of energy used to provide the same amount of light. The 11 watts versus the 60 watts describes the *difference* in energy use at a device-specific level. This change to an 11-watt light would typically be referred to as improving, enhancing, or increasing "energy efficiency." And that change might also be referred to as "energy conservation." Assuring that the light is turned off when the room is not occupied would typically be referred to as "energy conservation" but can also be referred to as improving, enhancing, or increasing "energy efficiency." These are both measures describing *change* in the way energy is used or *differences* between alternative ways energy can be used.

All energy-use decisions, taken together in an economy, result in the total energy used. *Energy intensity* of the economy is a summary statistic measuring the total energy used in the economy, and is not a measure of changes in energy use. Energy intensity of the economy is the ratio of (1) the total amount of energy used in all sectors of the economy to (2) the (inflation-adjusted) GDP. Equivalently, the "energy

---

5. LED is short for light-emitting diode.

productivity" of the economy is the ratio of (2) the GDP to (1) the total amount of energy used in all sectors of the economy.

Energy efficiency and energy productivity are simply mathematical inverses of each other. For example, in 2014 the energy intensity of the US economy was 6120 BTUs per dollar of GDP (2009 dollars). The energy productivity was $163.40 of GDP per million BTUs of energy used.[6] One percent less energy used for a fixed GDP would reduce the energy intensity by 1 percent and equivalently would increase the energy productivity by 1 percent.

In this book, I primarily use the term *energy intensity* and only secondarily use *energy productivity*. Energy intensity has been the more common measurement and accurately describes the concept.[7]

The combined impacts of all energy-efficiency or energy-conservation changes are reflected in the aggregate statistic, energy intensity. Energy intensity gives an indication of the economy-wide total of energy-efficiency enhancements but does not directly measure energy efficiency.

Energy intensity is calculated and reported by statistical agencies of the US government, in particular, the US Department of Energy, Energy Information Administration. Although some conventions are needed to calculate both total energy use[8] and GDP, these conventions

---

6. 1,000,000 BTUs/6120 BTUs per $ of GDP = $163.40 of GDP per million BTUs of energy used.

7. "Energy productivity" has been used partly as a marketing strategy to promote energy efficiency. See, for example, "Part of it [the term *energy productivity*] is messaging and talking about energy efficiency in a different way," says Nicole Steele, program manager of Policy and Research for the Alliance to Save Energy. Energy efficiency has a negative connotation of turning down a thermostat and wearing a sweater, she adds. "Now there's a more positive way to message energy efficiency—and [you can] use it to get more out of your business." See http://greentechadvocates.com/2013/02/14/goodbye-energy-efficiency-hello-energy-productivity/.

8. Conventions are needed particularly for power plants that generate electricity from hydro, geothermal, solar thermal, photovoltaic, and wind energy sources. For these plants, there is no generally accepted practice for measuring the thermal conversion rates. This book adopts the convention that the US Energy Information Administration uses. The EIA calculates a rate factor

remain the same over extended periods of time, so that energy intensity (or energy productivity) is measured and reported consistently over time. Given the conventions, there would be no disagreements about the measure of energy intensity.

Although energy intensity is an agreed-upon and objectively measured concept, the term *energy efficiency* is used differently by the various scientific disciplines[9] and thus can have different meanings among people. One precise definition comes from physics. Efficiency is a ratio: energy transferred to a useful form compared to the total energy initially supplied. But that definition, though precise, is too narrow a concept for energy policy discussions.

A broader concept, useful for economic policy analysis and used by the International Energy Agency, is "something is more energy efficient if it delivers more services for the same energy input, or the same services for less energy input."[10]

Although the latter concept is a good starting point for analysis of the energy system, one must be clear what is meant by "more services" or "the same services." If an automobile drives 100 miles using 5 gallons of gasoline, and another identically drives the same 100 miles using 8 gallons, one can easily say that the first is more energy efficient. But if one person uses a video connection to a distant meeting using very little energy and another flies to the same meeting using much more energy, the first may well be more energy efficient, even though the final services may be somewhat different—being at the meeting is somewhat different from participating in the meeting only by video.

---

that is equal to the annual average heat rate factor for fossil-fueled power plants in the United States (text quoted from US EIA, *Annual Energy Review*, Appendix B).

9. A good discussion of the many meanings comes from the Wuppertal Institute. See the web pages: http://wupperinst.org/uploads/tx_wupperinst/energy_efficiency_definition.pdf.

10. http://www.iea.org/topics/energyefficiency/

Extending the idea: if a company shifts the nature of the goods it produces, uses less energy, and increases its profits, one could say that the firm was becoming more energy efficient. For the definition of energy efficiency, the services would be "meeting" or "earning a profit."

More broadly extending the concept, if the US economy structurally shifts to using less energy but does not decrease GDP in doing so, it can be said to have become more energy efficient. For the definition of energy efficiency, the services would be "economic output," that is, "services" would be closely approximated by GDP. In this book, such changes are counted as energy-efficiency enhancements because they reduce the use of energy but do not reduce the overall value of goods and services available to the US economy.[11]

In energy efficiency, this book includes changes resulting from telecommuting to meetings, profitably producing less energy-intense outputs, shifts in modes of transport, product changes resulting from R&D targeted toward less energy use, changing cultural norms toward using less energy, and structural shifts in the economy, as long as these do not decrease the sum total value to all members of the economy.

Another concept is that of *energy conservation*, which also can have several different meanings. Widely used is the idea: "Energy conservation refers to reducing energy consumption through using less of an energy service."[12] This concept fits with the concept that something is

---

11. There is not broad agreement as to whether such structural shifts should be counted as energy-efficient changes, even though they reduce the use of energy and increase (or do not reduce) GDP. For example, in some studies, the Energy Efficiency and Renewable Energy office of the US Department of Energy excludes such structural shifts. Thus, although the energy intensity of the economy is an objective measure, different groups could differ in estimates of energy-efficiency impacts, based on whether they count or exclude structural shifts as energy efficiency. In addition, it may be difficult to determine whether particular choices are economically efficient. That knowledge normally can be inferred by understanding whether choices are made willingly or by understanding whether regulatory processes are conducted to require only economically efficient changes.

12. "Energy Conservation," Wikipedia, https://en.wikipedia.org/wiki/Energy_conservation.

energy efficient if it also delivers more services for the same energy input, or the same services for less energy input.

The concept would be simple if discussions were limited to a discrete action, such as turning off a light. But as in the discussion of the energy-efficiency concept, the definition of *energy service* is essential to discussion of the energy conservation concept. If *energy service* were defined consistently, energy-use reductions would be either energy conservation or energy efficiency, but not both, depending on whether there was a reduction of the service. The difficulty is that often the concept of *service* is implicitly used differently when discussing energy efficiency or energy conservation.

Another definition of energy conservation is "reduction in the amount of energy consumed in a process or system, or by an organization or society, through economy, elimination of waste, and rational use."[13] That second definition is essentially the same as the energy-efficiency definition. Under this definition, the concepts of energy efficiency and energy conservation are virtually identical. Under this definition *energy efficiency* and *energy conservation* would be synonymous.

Rather than confront the various, often contradictory, connotations of the term *energy conservation*, this book avoids using this term except when necessary, such as quoting other documents. Instead, this book focuses on *energy efficiency* and uses *energy intensity* as a way of indicating the amount of energy-efficiency enhancements that have occurred over time in the US economy.

Barriers to Energy Efficiency

Even with its many economic, environmental, and security benefits, numerous barriers inhibit full implementation of energy-efficient actions. Many energy options are not chosen, even though they appear to

---

13. "Energy Conservation," BusinessDictionary, http://www.businessdictionary.com /definition/energy-conservation.html.

be desirable to the individual or organization making the decision. These are often discussed informally as "unpicked fruit" or "fruit rotting on the tree." More formally, the failure to choose optimally is referred to as an "energy efficiency gap." It is not the purpose of this book to examine these barriers in any depth. But they can be summarized in terms of market failures, institutional barriers, and behavioral issues, all typically leading to underutilization of energy-efficiency options. Some barriers are listed in Table 1.1.

Much literature has been aimed at identifying methods for addressing these various barriers,[14] and interested readers can find much valuable material. Suffice it to say, however, that there is good news and bad news. The bad news is that fixing these issues can be very hard. The good news is that there are further opportunities to enhance energy efficiency based upon overcoming barriers. And as will be seen from the discussions in this book and in the other sources that have identified the issue of energy-efficiency barriers, many policy and organizational changes already being implemented are designed to overcome these market, institutional, and behavioral barriers.

There is other good news, based on the history of energy-efficiency progress in the United States. Even though many energy-efficiency options remain, the energy-efficiency enhancements that have already been broadly implemented are ubiquitous throughout the US economy. This history shows that, as stated above, energy-efficiency enhancements since the energy crisis of 1973–74 have had more beneficial impacts on US energy security and on the environment than all of the increases in domestic production of oil, gas, coal, geothermal energy, nuclear power, solar power, wind power, and biofuels combined.

---

14. For example, see Marilyn A. Brown, "Market Failures and Barriers as a Basis for Clean Energy Policies," *Energy Policy* 29 (2001): 1197–1207. Or see K. Gillingham and J. Sweeney, "Barriers to Implementing Low Carbon Technologies," *Climate Change Economics*, 3, no. 4 (2012): 1–25. doi: 10.1142/S2010007812500194.

MARKET FAILURES
Poor information about energy use of equipment/appliances
Principal/agent problems (split incentive problems)
Limited time rental or ownership
Transaction costs in efficiency markets
Externalities: in use; R&D spillovers
R&D spillovers
Learning-by-doing spillovers
Technology lock-in
Market power of makers of existing technologies
Network externality: complementary products requiring large nonrecoverable investments

INSTITUTIONAL BARRIERS
Fragmented structure of construction industry
Limited modeling tools for building design
Corporation organization (energy as overhead)
Limited energy information and control systems
Distortionary regulatory and fiscal policies (e.g., obsolete building codes)
Local governmental land use institutions (e.g., zoning or transport decisions)

BEHAVIORAL ISSUES
Low salience of energy issues
Transaction costs: efficiency upgrades, determining optimal purchases
Limited cognitive skills
Poor information about electricity prices
Poor information about equipment/appliances energy use
Incentives and priorities for managers
Limited feedback: energy choices to outcomes

Table 1.1. Some Barriers to Energy-Use Optimality

The fundamental importance of energy efficiency may come as a surprise because one cannot easily observe what energy use would have been without the greater efficiencies. Most energy-efficiency changes are rather invisible to outsiders (and often to insiders), and thus most improvements are unrecognized. Much of the progress has been the result of cumulative small changes broadly distributed throughout the economy. Such cumulative, small, broadly distributed changes together have greatly reduced the energy intensity of the US economy.

We now turn to examples of energy-efficient changes in the United States that have been broadly implemented since the oil crisis of 1973–74.

2

Energy Efficiency Is All Around Us

During the Post–World War II years until 1973 in the United States, very little attention was paid to the use of energy. But in 1973 the world of energy was fundamentally altered. As a result of the US support of Israel during the Yom Kippur War, the Arab OPEC[1] members initiated an oil embargo against the United States and several other countries. Of more lasting impact, the Arab OPEC members reduced their exports of oil, and as a result the world price of oil tripled virtually overnight.

The 1973–74 energy crisis brought attention to energy in both the public and private sectors and increased most energy prices. The crisis motivated an intense time of government policy innovation at both the state and national levels. It led businesses to change energy-related practices and to pay attention to energy use in their innovation. It motivated existing and newly created non-governmental organizations to focus attention on energy. And it encouraged individuals to pay attention to energy costs in their own lives. These factors together led to many changes that began a slow cumulative process of energy-efficiency improvements.

Most energy-efficiency changes taken alone had relatively little impact on the total energy use in the US economy. Except for increased fuel economy of cars and trucks (to be discussed later), each change had a quantitatively small percentage impact on the total energy use in the United States. However, these relatively small changes were broadly

---

1. Organization of Petroleum Exporting Nations.

distributed throughout the economy and collectively had a profound impact on US energy consumption.

This book takes 1973 and the pre-crisis years as a comparison period to show energy-efficiency enhancements throughout the United States. This chapter describes some of the specific changes in technologies, technology diffusion, and business practices that have occurred between then and now. These changes are chosen as important examples of some specific energy-efficiency enhancements. But there has been no attempt to be comprehensive. Many other changes have occurred, too numerous to discuss here.

New or Improved Technologies

Easiest to see are those energy-using technologies that are ubiquitous in homes, offices, industrial facilities, roadways, and in the air. The following paragraphs summarize some energy-efficiency improvements in these technologies that are all around us.

Lighting

"Revolution" is the best way to characterize changes in indoor and outdoor lighting. Artificial lighting is everywhere in the economy, in homes, stores, offices, hospitals, schools, factories, airplanes, cars, street lighting, and traffic lights. Everywhere.

Lighting technology in 2016 has radically improved in energy use from the technology of 1973. In 1973, most indoor lighting in the residential sector was based on incandescent bulbs. The primary source of artificial light in most homes was 60-watt, 75-watt, or 100-watt incandescent bulbs. Even night lights were incandescent. Exterior lighting also included incandescent floodlights or spotlights, typically drawing more than 100 watts of electricity each. In 1973, industrial and commercial-sector lighting was typically some combination of incandescent bulbs and fluorescent tubes. Fluorescent tubes used less electricity for the given amount of light but provided a harsher light quality than incandescent bulbs. A 1973-generation office light was typically a lensed rectangular light fixture in the ceiling, using four fluorescent tubes and magnetic ballasts, and consuming about 200 watts.

Soon after the energy crisis, a new form of fluorescent light, the compact fluorescent light (CFL), was invented. Designed to reduce electricity use, CFLs could fit into standard light fixtures designed for incandescent bulbs. The spiral CFL, although invented at General Electric[2] in 1975, was not produced by GE. However, in 1980, Philips introduced a screw-in CFL; in 1985, Osram offered another for sale. By the 1990s, a wide variety of US-produced and Chinese-produced lamps were broadly commercially available.

Compact Fluorescent Light (CFL)

Source: ID 16711910 © Alexmax/Dreamstime.com

Although originally CFLs were very expensive relative to incandescent lamps and were adopted only slowly at the start, continued R&D and production experience, led to sharp cost reductions.[3] At the time of this

---

2. http://americanhistory.si.edu/lighting/20thcent/invent20.htm - in4

3. Increased sales, which led to increased production, were spurred by state public utility commissions that required energy-efficiency program administrators (usually utilities) to subsidize costs with both upstream rebates to manufacturers and direct rebates to downstream individual and business purchases.

writing, a high-quality CFL manufactured by Philips can be purchased at retail online or in stores for about two dollars each. A CFL, providing as much light as a 60-watt incandescent light, uses only 13 watts, roughly one-fifth as much electricity as the comparable incandescent light.

But CFLs were not perfect substitutes for incandescent lights. The early CFLs provided harsh light; new ones have better quality. Early ones (and many current ones) were not dimmable. Most are dim when they originally turn on and take one to three minutes to reach full brightness.

More recently—after the turn of the century—even better energy-efficient lights were commercialized, the light-emitting diode (LED).[4] LEDs provide very high-quality light, turn on instantly, are dimmable, and use even less electricity than CFLs. A high-quality LED providing as much light as a 60-watt incandescent light uses 11 watts, roughly one-fifth as much electricity. LEDs are more expensive than CFLs but have a far longer life: typical LEDs guarantee a long life of approximately 50,000 hours—six years of continuous use.

Light Emitting Diode (LED)

Source: Photo by J. Navarrette

---

4. The LED lamp is based on semiconductors that emit light when subjected to a voltage. Phosphors coating the semiconductor convert that light into either cool white or warm white light.

Modern LEDs are now supplanting CFLs. Manufacturers include Cree and Philips Lumileds in the United States plus many manufacturers from other countries, particularly China. At the time of this writing, LED lights manufactured by Cree can be purchased at retail online or in stores for about eight dollars each. Others are available at even lower prices. And just as happened with CFLs, state utility commissions are requiring that the utility customer–funded lighting efficiency programs now focus on LEDs rather than CFLs, and are providing millions of dollars of upstream and downstream subsidies to lower costs to the ultimate end-use buyer.

Either a CFL or an LED saves the consumer more money than the cost of the LED or CFL over its lifetime.[5] And importantly, large-scale use of efficient lighting avoids building new, more expensive electric generation and transmission infrastructure, saving billions of dollars throughout utility systems and the US economy.

Other energy-efficient lighting technologies are less obvious. Traditionally, fluorescent lights used magnetic ballasts[6] that limited the operation of fluorescent lights to 60 cycles per second. DOE-sponsored R&D led to the development and commercialization of electronic ballasts.[7] Electronic ballasts now operate fluorescent lights at much higher

---

5. For example, if a light were used three hours a day, or 1095 hours per year, a 60-watt bulb would use 65.7 KWhr per year; a 13-watt CFL would use 14.2 KWhr; and an 11-watt LED would use 12.0 KWhr. If the consumer paid $0.13 per KWhr of electricity, the annual cost would be $8.54 for the 60-watt bulb, $1.85 for the CFL, and $1.57 for the LED. Each CFL that replaced a 60-watt incandescent bulb would save $6.69 per year; each LED would save $6.97. Over a few years, this savings would more than surpass the $2 or $8 purchase cost. CFLs would typically last many years. An LED having a 50,000-hour lifetime could be used, at least in theory, for 46 years!

6. The ballast regulates the current to the CFL, providing sufficient voltage to create an arc between the two ends of the tube, but limiting the current once the arc is established.

7. *Energy Research at DOE, Was It Worth It? Energy Efficiency and Fossil Energy Research 1978 to 2000*, Committee on Benefits of DOE R&D on Energy Efficiency and Fossil Energy, Board on Energy and Environmental Systems, Division on Engineering and Physical Sciences, National Research Council, p. 104.

frequencies, thus increasing their efficiency by at least 10 percent. Electronic ballasts are now the dominant ballast technologies for fluorescent lights.

With the lighting revolution, homes and offices now can be far more energy efficient. In contrast to 1973, in 2016 homes range from some with almost all incandescent lighting to others with almost all LED lights. Today's LED office fixture consumes about 35 watts, not 200 watts, and provides better quality light.

The invention and commercialization of the LED has fundamentally impacted the efficacy of light sources in virtually all general lighting uses, as summarized in Table 2.1.

Although Table 2.1 shows the fundamental impact of LEDs in their various applications, there has not yet been 100 percent penetration in these various uses. But a nearly 100 percent ultimate penetration is likely, thus we can expect a continuation of the ongoing revolution in light sources. A continued focus is needed because the actual penetration of LEDs in the lighting area, though growing, is still low. For economic and environmental reasons, particularly climate change, accelerated replacement of incandescent and CFLs with LEDs continues to be a major focus by utilities in their efficiency programs and by regulators.

Not only did light sources become more energy efficient, but other innovations in lighting have been similarly important. For example, office lights now can be readily turned off when not needed. In 1973, typically all offices on one floor were controlled by a single master switch. When one office needed light, all were lighted. And for many office buildings, the lights remained on all night, often 24 hours a day, 365 days per year. Now, typically, individual offices have lighting controls; lights can be turned on without turning on lights in other offices, so the others can stay off.

APPLICATION	1973		2015		
	Light Source	Efficacy (LPW)	Light Source	Efficacy (LPW)	Watts/ Lumen Reduction
Lightbulb	Typical 60-watt incandescent (A-19)	14	LED bulb equivalent (A-19)	84	83%
Cobrahead Streetlight	High-pressure sodium	48	LED	93	48%
High Day Industrial	400-watt metal halide (14K lumens)	31	213-watt LED (18K lumens)	85	64%
Office Recessed 2×4 Luminaire	40-watt, T12 fluorescent	60	2×4 recessed LED luminaire	115	48%
Kitchen Down Light	5-inch diameter, 65-watt incandescent (BR40)	10	5-inch diameter, 12-watt LED (BR40)	67	85%
Track Lighting	2.5-inch diameter, 45-watt spot incandescent (R20)	9	2.5-inch diameter, 5-watt LED (R20)	65	87%

Table 2.1. Changing Light Efficacy: Lumens per Watt (LPW): 2015 vs. 1973
Source: Finelite, Inc.

In addition, in sharp contrast to 1973, lighting in office buildings, some homes, and in some retail spaces is controlled by motion-sensing/light-sensing switches that turn off lights if either there is no motion in the room or there is sufficient daylight. Such switches may turn on the lights automatically when a person enters a darkened room. Parents, who may frequently remind their children to turn off lights when they leave rooms, don't themselves have to remember to do so when they leave *their* offices!

Motion Sensing or On/Off Light Switch, Single Pole

Source: ID 46468569 © Kostyantin/Dreamstime.com

Motion sensing has extended to some stores, particularly those open late at night. In these stores, motion sensors detect when a customer is starting down an aisle, and turn on light above that aisle. Lights are automatically turned off once there are no customers in that aisle. For example, Walmart has used this approach.

Although motion sensors had been invented well before the 1973 energy crisis, the improvement in the sensitivity of the detection and the reduction in cost allowed motion-sensing/light-sensing switches to be broadly installed.

Refrigeration

Virtually every American home, restaurant, bar, grocery store, or other purveyor of perishable foods has at least one refrigerator or refrigeration unit, and many have more than one. Prior to the 1973 energy crisis, the average size of new home refrigerators was increasing annually, and the electricity use per refrigerator was increasing even more rapidly. Although refrigerator size has not decreased since 1973, the trend of increasing electricity use was sharply reversed soon after the energy crisis, thanks to a combination of technology advances, performance

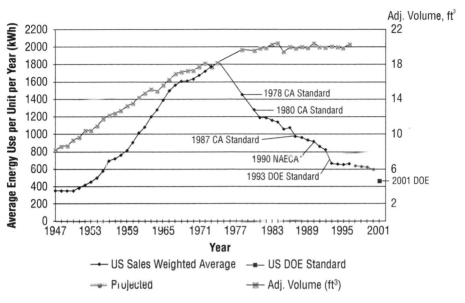

Figure 2.1. Energy Efficiency of Refrigerators

Source: Goldstein and Geller (1998)

standards, utility incentive programs, ENERGY STAR labeling, federal tax credits, and the Golden Carrot initiative.[8]

Two graphs show these changes. Figure 2.1 shows trends of refrigerator size and annual electricity use for those sold before 1973 and those sold between 1973 and 2001.[9] The second,[10] Figure 2.2, provides

---

8. Utilities in California and the Northwest, working with their state commission regulators, organized and funded the Super Efficient Refrigerator Program (SERP), which featured a $30 million bid competitively awarded to the refrigerator manufacturer that could develop, distribute, promote, and sell the most energy-efficient, CFC-free refrigerator/freezer in the most cost-effective manner possible. SERP is discussed more fully in Chapter 6.

9. From *Energy Research at DOE, Was It Worth It?*, p. 97. Original data from Goldstein and Geller, 1998.

10. Source: www.appliance-standards.org/sites/default/files/Refrigerator_Graph_Nov_2015.pdf.

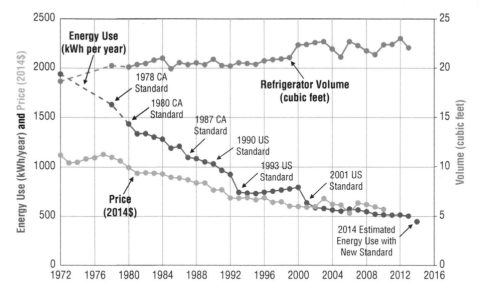

Figure 2.2. New Refrigerator Energy Use, Volume, and Price Trends

Data source: Association of Home Appliance Manufacturers (AHAM) for energy consumption and volume, US Census Bureau for price.
Graph source: www.appliance-standards.org/sites/default/Refrigerator_Graph_Nov_2015.pdf

Notes: a. Data include standard-size and compact refrigerators.
 b. Energy consumption and volume data reflects the DOE test procedure published in 2010.
 c. Volume is adjusted volume, which is equal to fresh food volume + 1.76 * freezer volume.
 d. Prices represent the manufacturer selling price (e.g., excluding retailer markups) and reflect products manufactured in the United States.

complete data for refrigerators sold between 1973 and 2013. In addition, Figure 2.2 includes information about average refrigerator prices.

By 1980, the electricity used for the average new refrigerator had been reduced by about one-third and by 2013, by about three-quarters of its 1973 level, even though refrigerators had increased in average size and decreased in (inflation-adjusted) price.

Cars and Light Trucks

Most personal mobility is based on cars and light-duty trucks (such as SUVs, minivans, and pickup trucks.) Prior to the crisis, fuel economy of cars and trucks had been gradually declining. In 1973, the average fuel efficiency of all light-duty vehicles on the road was about 12.5 miles per gallon (mpg); the average light-duty vehicle used 8 gallons per hundred miles (gphm) of driving.[11]

The energy crisis of 1973–74 motivated Congress and the Ford Administration to pass and sign into law the Energy Policy and Conservation Act (EPCA) of 1975. Among its many provisions, EPCA included corporate average fuel efficiency (CAFE) standards, standards placed on each manufacturer limiting the average of the fuel efficiency—mpg—of new cars and light-duty trucks it sold in the United States each model year.[12] Subsequently, numerous efforts to strengthen those standards were blocked over a 30-year period until Congress passed the Energy Independence and Security Act of 2007, allowing those standards to be tightened over time. President Obama, in 2012, increased the required fuel economy to 54.5 mpg by 2025 for the average of cars and light-duty trucks.

CAFE standards were designed to reduce the amount of gasoline used by light-duty vehicles—cars and light trucks—while allowing automobile manufacturers to sell a wide range of different vehicle models and giving the manufacturers flexibility in designing and marketing their products. The CAFE standards for passenger cars first came into effect in 1978; standards for trucks came into effect in 1979.

By 1985, fuel economy of light-duty vehicles had increased to 17 mpg, so on average a new vehicle consumed 5.9 gallons of fuel per hundred miles of driving. And as of 2013, fuel consumption of new vehicles had dropped to about 4.5 gphm of driving, 57 percent of energy use per mile of driving in 1973.

---

11. To convert from miles per gallon (mpg) to gallons per 100 miles (gphm), divide 100 by mpg to get gphm.

12. More precisely, in a given model year the standard was a lower limit on the production-weighted harmonic mean fuel economy, expressed in mpg. Mathematically, this was the same as placing an upper constraint on the production-weighted average of gallons per mile.

Figure 2.3, developed by Michael Sivak and Brandon Schoettle[13] shows fuel economy on-the-road data for passenger cars, light-duty trucks, the average of all light-duty vehicles, and other trucks. For ease of reference, a vertical line at the year 1973 has been added to the Sivak/Schoettle graph. The graph shows the continual decrease in mpg before 1973 for three averages: cars, all trucks, and all vehicles. The graph also shows that the average mpg of four groupings (cars, all trucks, all light-duty vehicles, all vehicles) remained roughly constant from the time of the crisis until the CAFE standards came into effect. But once the CAFE standards were in place, the four mpg averages began increasing rapidly until 1991, when the progress began slowing down. It was not until 2006, when gasoline again became more expensive, that the average fuel economy of all light-duty vehicles again started increasing.

The CAFE standards specified higher average mpg of cars than of trucks, recognizing that trucks normally were heavier than cars. This difference is shown in Figure 2.3, with the mpg of cars always larger than that of light trucks. In 1975, when the CAFE standards were first passed, trucks were typically used for work, and cars were typically used for personal mobility. However, over time this generalization crumbled, with the increased sales of SUVs and mini-vans, both of which were categorized as trucks, as well as the growth of pickup trucks for personal transportation.

As seen in Figure 2.3, the fuel economy average of all light-duty vehicles in some years rose less rapidly and others more rapidly than did the fuel economy of cars or light trucks. When the market share of cars relative to that of trucks increased, the average fuel economy of all light-

13. M. Sivak and B. Schoettle (2015) "On-Road Fuel Economy of Vehicles in the United States: 1923–2013," Report No. 2015-25 (Ann Arbor: The University of Michigan Transportation Research Institute). Available at http://deepblue.lib.umich.edu/bitstream /handle/2027.42/115486/103218.pdf.

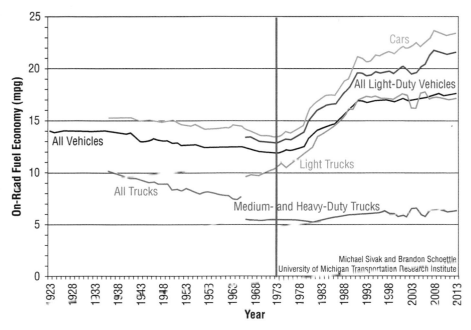

Figure 2.3. Historical On-Road Fuel Economy from 1923 to 2013

Source: Michael Sivak and Brandon Schoettle (2015)

duty vehicles increased more rapidly (e.g., beginning[14] in 2006); the reverse was true when the relative market share of cars decreased (e.g., from 1995 to 2006.)

The graph shows that medium- and heavy-duty trucks kept roughly the same fuel economy for the last 45 years. CAFE standards applied only to light-duty vehicles and not to medium-duty or heavy-duty trucks, so the regulatory push of CAFE standards did not apply to these trucks.[15]

---

14. People started buying relatively fewer trucks, such as SUVs, and relatively more cars when gasoline prices started rising. This shift in the market shares of cars versus trucks led to the average mpg of all light-duty vehicles increasing more rapidly than the mpg of either trucks or cars.

15. It is an open question as to whether efficiency standards would have increased or decreased the economic efficiency had they been applied to these trucks. The trucks were

However, fuel-efficiency gains did not come automatically. Meeting the standards required manufacturers to make many changes, some costly. Manufacturers had to redesign vehicles to meet the higher fuel economy standards. Technologies embedded in the engines, drivetrain, and automobile bodies were redesigned to require less energy. Automatic transmissions were improved. Cars were designed to be lighter, for example, by use of aluminum engines. The body designs were engineered so cars would be shorter and thus lighter. The front end of the vehicles and the back end were designed to be smaller, keeping the passenger compartment roughly the same size. Manufacturers improved the vehicle aerodynamic design in order to reduce air resistance. Better tires had less rolling resistance and therefore increased fuel economy. Hybrid electric vehicles were introduced. And manufacturers altered their marketing and advertising to encourage consumers toward buying smaller, more fuel-efficient vehicles.

1973 Chevy Impala

Source: http://carphotos.cardomain.com/ride_images/4/449/2261/38621130001
_original.jpg

---

typically operated by entities that had an interest in purchasing the trucks that had the lowest discounted life cycle cost. Whether the overall system was economically optimal has not been resolved.

2014 Chevy Impala

Source: http://www.chevrolet.com/content/dam/Chevrolet/northamerica/usa/nscwebsite/en/Home
/Vehicles/Cars/2016_Impala/Model_Overview/01_images/2016-chevrolet-impala-full-size-sedan-build
-your-own-647x200-001.jpg

The net result has been high performance vehicles that use roughly half as much gasoline per mile of driving as did the vehicles sold before the energy crisis.

These gains were not free and the upfront cost of vehicles increased. The question is whether the gains in fuel economy were worth the cost, that is, whether the gains were truly energy efficient— economically efficient reductions in energy use—so that the increased upfront cost was offset by fuel or other savings over the lifetime of the car. The evidence is that these changes in fact improved economic efficiency, that the cost of the more fuel-efficient cars was smaller than the value to car buyers of the gasoline savings, and that even higher fuel efficiency would be desirable.[16] The relatively small increase in new purchase prices was more than offset by the savings in gasoline costs.

---

16. In 2003, the National Research Council addressed this question and concluded that the benefits exceeded the costs. It also examined whether even greater increases in fuel economy would be economically justified. Among its many conclusions, it found "the calculation indicates that the cost-efficient increase in (average) fuel economy for automobiles could be increased by 12 percent for subcompacts and up to 27 percent for large passenger cars. For light-duty trucks, an increase of 25–42 percent (average) is calculated, with the larger increases for larger vehicles." See *Effectiveness and Impact of Corporate Average Fuel Economy (CAFE) Standards*, Committee on the Effectiveness and Impact of Corporate Average Fuel Economy (CAFE) Standards, National Research Council, 2003, p. 66.

In addition to the net benefits accruing to new car buyers, there is a financial benefit to the United States as a whole from increased fuel efficiency of cars and trucks. Decreases in the consumption or increases in the production of oil in the US place downward pressure on the price of imported oil. Because the United States historically has been a large importer of oil, reductions in the world price have led to terms-of-trade economic benefits. Individual consumers normally cannot be expected to take into account this impact on the overall US economy and thus often choose vehicles that have lower mpg than is optimal for the US economy as a whole. This behavioral tendency reinforces the economic value to the United States of more fuel-efficient vehicles.

What is apparent in retrospect was controversial at the time the CAFE standards were being debated in Congress. At that time, many opponents of the CAFE standards suggested that fuel-efficient vehicles by necessity would have poor performance and that low acceleration rates of fuel-efficient vehicles would lead to traffic accidents. Cars would not be able to safely accelerate onto freeways, it was asserted. In reality, performance was not compromised. As the CAFE standards came into effect, the average acceleration time—0 to 60 miles per hour—did not change significantly, as shown in Figure 2.4; it decreased in later years.[17]

The changing fuel efficiency of vehicles had little impact on mobility, as indicated by the total vehicle miles of travel in the United States.[18]

17. It is possible that average acceleration time would have decreased more rapidly, absent CAFE standards. But it is dubious that the failure to reduce acceleration time as much has led to traffic accidents. Data from US Environmental Protection Agency, "Light-Duty Automotive Technology, Carbon Dioxide Emissions, and Fuel Economy Trends: 1975 Through 2013," p. 4, Table 2.1. https://www.fueleconomy.gov/feg/pdfs/420r13011_EPA_LD_FE_2013_TRENDS .pdf.

18. Data source: US Department of Transportation, Federal Highway Administration, Policy and Governmental Affairs, Office of Highway Policy Information, "Highway Statistics 2013." http://www.fhwa.dot.gov/policyinformation/statistics/2013/vmt421c.cfm. Vehicle miles traveled is an imperfect measure of mobility, since it does not describe whether load factors have

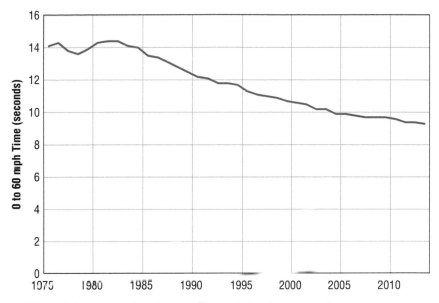

Figure 2.4. Average AccelerationTime of New Passenger Cars

Source: Data from US Environmental Protection Agency, "Light-Duty Automotive Technology, Carbon Dioxide Emissions, and Fuel Economy Trends: 1975 Through 2013," p. 4, Table 2.1

As shown in Figure 2.5, total vehicle miles was reduced a small amount from the pre-1973 trend but continued to grow at almost the same rate until 2006. Over the last eight years, however, there has been relatively little growth in total vehicle miles traveled (VMT). The various causes of that hiatus in travel are still uncertain, but at least in part seem to be related to high oil prices and to demographic and behavioral shifts.[19]

---

changed. For example, more car-pooling could increase the ratio of passenger-miles to vehicle miles.

19. This is a very active area of current research. A 2015 paper presented at the annual Transportation Research Board meetings documented that this phenomenon started at the state level as early as 1992, and by 2011, 48 of 50 states had peaked in miles traveled per capita (see Garceau, Atkinson-Palombo, and Garrick, 2015). A 2013 report by the University of Wisconsin's State Smart Transportation Initiative suggests several overlapping factors have affected VMT. These include a weakened influence of the following factors: "large-scale highway construction,

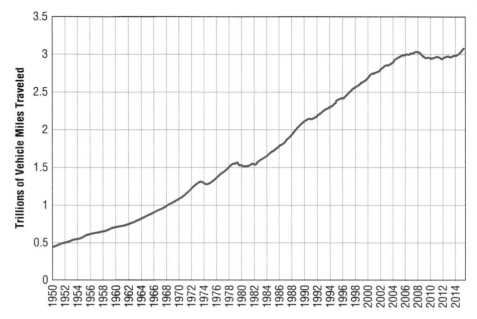

Figure 2.5. Vehicle Miles Traveled (trillions): All US Roads

Source: US Department of Transportation. Federal Highway Administration. Policy and Governmental Affairs; Office of Highway Policy Information "Highway Statistics 2013"

Despite the growth in the amount of travel, the reduction in the fuel use per mile of travel for light-duty vehicles is a key factor underlying the reduced energy-use growth rate in the transportation sector. This conclusion is illustrated in Figure 2.6, which shows estimates[20] of

---

women entering the workforce, a large baby boom population, growing incomes, rising automobile ownership, and restrictions on compact, mixed-use development" (Chris McCahill and Chris Spahr, *VMT Inflection Point: Factors Affecting 21st Century Travel*, State Smart Transportation Initiative, September 2013, http://www.ssti.us/2013/09/vmt-inflection-point-factors-affecting-21st-century-travel-ssti-2013/). It also includes "limits on automobile travel and changing preferences for non-automobile modes." In addition, the Federal Highway Administration method of calculating data for 2007 and later changed from that of previous years.

20. Basic data are from *Transportation Data Book*, Oak Ridge National Laboratory, http://cta.ornl.gov/data/spreadsheets.shtml.

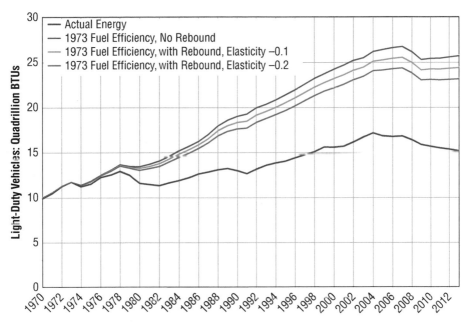

Figure 2.6. Energy Used by Cars and Light Trucks: Actual and Three Limited-Energy-Efficiency Benchmarks

the fuel use for cars and light-duty trucks, measured in terms of qua-drillion BTUs, in actuality and in hypothetical cases in which average fuel economy of light-duty vehicles remained at the 1973 level.

Three different hypothetical cases are shown. In the first case, shown in red, the calculation is based on an assumption that under the limited-energy-efficiency benchmark, vehicle miles traveled would have been the same as the actual vehicle miles traveled. In the other two cases, shown in green and purple, a rebound effect[21] has been

---

21. The rebound effect is the phenomenon that if miles-per-gallon fuel use had been lower, the cost per mile of driving would have been equivalently higher, and people would have driven fewer miles. These calculations use a constant elasticity of demand function with elasticities of either –0.1 or –0.2 applied to the cost per mile of driving, which is inversely proportional to the

included. These two cases are based on an assumption that, in response to the higher cost per mile of driving under the limited-energy-efficiency benchmark, there would have been a reduction in the vehicle miles traveled. These calculations are based on a price elasticity of demand of either –0.1 or –0.2.

Figure 2.6 shows that by 2013 there was an estimated 8- to 10-quadrillion BTU reduction in energy use between any of the three limited-energy-efficiency-gains benchmarks and the actual energy use for light-duty vehicles.

There have been debates about whether the average fuel economy of new cars and trucks has been the result of CAFE standards or increased gasoline prices. This issue has become more important now that the federal government has tightened fuel economy standards, and gasoline prices have fallen sharply with the decline in crude oil prices.

In the late 1970s and early 1980s, the average fuel efficiency of new cars and trucks followed the CAFE standards closely. The rate of fuel economy increases was dramatic. For this time, it appears that the CAFE standards were the constraint, and the high gasoline prices made it easier for auto makers to meet the CAFE standards. Consumers were looking for more-fuel-efficient vehicles. Once oil prices dropped in 1985, high gasoline prices were no longer a factor, but the available technology had improved greatly, and the manufacturers had designed and marketed high-fuel-economy vehicles. Manufacturers were still able to meet and often to exceed the CAFE standards. Consumers had accepted the newly designed fuel-efficient vehicles, and the average fuel economy of cars exceeded the CAFE standards and kept growing while the standards remained constant. In the years after 2004,

average mpg. Rebound effects may be relevant in other areas of energy consumption as well, but the aggregate data on energy intensity imply that rebound effects have been dominated by the many factors leading to energy efficiency enhancements.

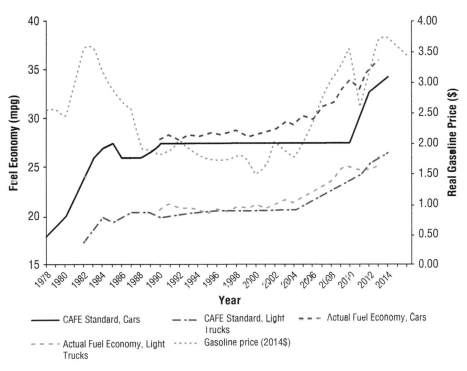

Figure 2.7. Fuel Economy Standards and Actual Fuel Economy of Cars and Trucks by Year, Plotted against Real Gasoline Prices in 2014 Dollars

Source: Figure copied from http://www.nap.edu/read/21744/chapter/11#311

gasoline prices again started rising rapidly, and average fuel efficiency increased even more rapidly. However, for trucks, average economy followed the CAFE standards closely. These data for new cars and trucks are shown in Figure 2.7, from a 2015 study of the National Research Council of the National Academies.[22]

22. Figure from *Cost, Effectiveness, And Deployment Of Fuel Economy Technologies For Light-Duty Vehicles,* Committee on the Assessment of Technologies for Improving Fuel Economy of Light-Duty Vehicles, Phase 2. Board on Energy and Environmental Systems, Division on Engineering and Physical Sciences, National Research Council of the National Academies, The National Academies Press, Washington, DC, 2015, p. 311. Note from the original source: "All

How much of the fuel economy increases in the 1990s and later were the result of the CAFE standards and how much were the result of consumer choice is not clear, given the further improvement of technologies, the manufacturers' existing portfolios of high-fuel-economy vehicles, and the consumer response to increasing gasoline price.

Although there have been attempts to disentangle the relative roles, the original National Research Council study was not able to determine the complete mix. It stated:

> While attempts have been made to estimate the relative contributions of fuel prices and the CAFE standards to this improvement . . . the committee does not believe that responsibility can be definitively allocated. Clearly, both were important, as were efforts by carmakers to take weight out of cars as a cost-saving measure. *CAFE standards have played a leading role in preventing fuel economy levels from dropping as fuel prices declined in the 1990s.*[23] [emphasis added]

And the 2015 study—*Cost, Effectiveness, and Deployment Of Fuel Economy Technologies For Light-Duty Vehicles*—primarily looked forward and did not come to any definitive conclusions about the past.

Regardless of the relative importance of the two factors, data monitored by Michael Sivak and Brandon Schoettle[24] show that the CAFE standards and relatively high gasoline prices (until recently) together have continued to keep high fuel efficiency of new vehicles. Average mpg of new vehicles has decreased by a small amount now that gasoline prices have decreased substantially. Figure 2.8 shows recent data on the monthly and model-year average sales-weighted average fuel economy.

---

fuel economy standard and actual values are the certification fuel economy values. Source: DOT (2014); BTS (2014); EIA (2014)."

23. "Effectiveness and Impact of Corporate Average Fuel Economy (CAFE) Standards, Committee on the Effectiveness and Impact of Corporate Average Fuel Economy (CAFE) Standards, Board on Energy and Environmental Systems, National Research Council, 2003, p. 15. James Sweeney was a member of that committee.

24. http://www.umich.edu/~umtriswt/EDI_sales-weighted-mpg.html

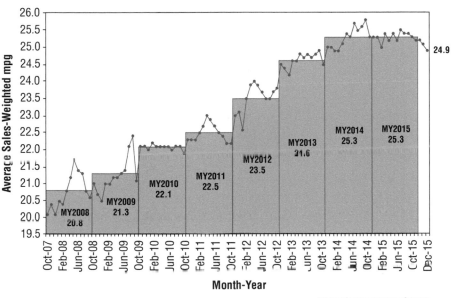

Figure 2.8. Average Sales-Weighted MPG of New Vehicles

Source: Michael Sivak and Brandon Schoettle; http://www.umich.edu/~umtriswt/EDI_sales-weighted-mpg.html

New vehicles for model year 2008 averaged 20.8 mpg (4.8 gphm). For model year 2015, average fuel efficiency increased to 25.0 average mpg (4.0 gphm).

Looking forward, one may ask whether the pattern of efficiency gains will continue. Gasoline prices have recently dropped substantially. As indicated above, President Obama, in 2012, increased the required fuel economy to 54.5 mpg for cars and light-duty trucks by 2025. The current low gasoline prices will make it very difficult for manufacturers to meet these new CAFE standards. Based on the history of manufacturer responses to CAFE standards, it is likely that whatever the CAFE standards are (within reason), they will be met, even if gasoline prices do remain low.

But specific standards are not set legislatively. CAFE standards are set and enforced by the US Department of Transportation, National Highway Traffic and Safety Administration. The US Environmental Protection Agency sets the greenhouse gas standards, generally consistent with CAFE standards. And these agencies are run by presidential appointees. Thus the next US president has the power to revise those standards, either upward or downward, just as President Obama has done. Whether the pattern of fuel efficiency gains will continue depends greatly on decisions to be made by future US presidents as well as on technological options and economic incentives.

We turn now to another transportation technology—aircraft.

Aircraft

In contrast to cars and trucks, airplanes were improving in efficiency well before the energy crisis. Efficiency has continued to improve, up through the latest airplanes made commercially available by Boeing and Airbus.

A large portion of these reductions were the result of changes in aircraft technology, including the engines and the aircraft body. As shown in Figure 2.9, an IPCC special report in 1997 showed that aircraft engines used about 40 percent less fuel (per unit of thrust) in 1997 than in 1960. The same report showed that aircraft fuel burn per seat showed further gains, so that between the years 1960 and 1997, aircraft fuel burn per seat mile for new airplanes was reduced 70 percent. Further gains continued after that time.[25]

Although the IPCC report uses 1960 as a base, its data suggests that fuel use per seat mile of newly introduced airplanes from 1973 until

---

25. Subsequently, a 2005 report conducted by three Netherlands researchers from the Nationaal Lucht- en Ruimtevaartlaboratorium (National Aerospace Laboratory NLR) extended and critiqued the IPCC report. It confirmed the report for jet airlines but cautioned that "the last piston-powered aircraft were as fuel-efficient as the current average jet," http://www .transportenvironment.org/sites/te/files/media/2005-12_nlr_aviation_fuel_efficiency.pdf.

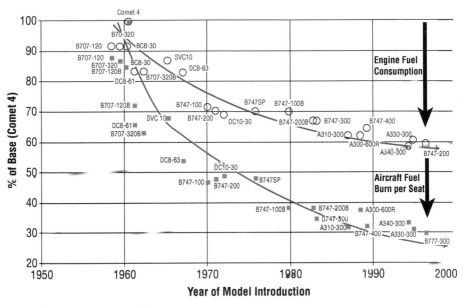

Figure 2.9. Energy-Efficiency Gains for Airplanes

Source: Aviation and the Global Atmosphere, IPPC Special Report 1997

1997 decreased by about 40 percent and would be expected to decrease even more by 2014.

These energy-efficiency gains were the result of many engineering decisions that included energy efficiency as an important design characteristic. Airline makers, such as Boeing and Airbus, understood well that their customers, the airlines, continued to evaluate fuel costs of the airplanes they might acquire. Saving fuel has been a continuous objective, because fuel has accounted for 12.5–32.5 percent of total airline costs.[26] Market forces—air passengers choosing airlines, airlines choosing airplanes, makers of aircraft choosing jet engines, and other

26. Source: J. J. Lee, S. P. Lukachko, I. A. Waitz, and A. Schäfer (2001) "Historical and Future Trends in Aircraft Performance, Cost, and Emissions," *Annual Review of Energy and the Environment* 2001, 26: 167–200.

technology improvements—have continued to motivate energy-efficiency gains in airplanes as a long-term cost-reduction strategy.[27]

One example, winglets—relatively small vertical wing extensions that reduce wing-tip drag—can reduce fuel use per mile by several percentage points while airplanes are cruising.

The development of winglets was a direct response to the energy crisis; winglets were developed explicitly for energy efficiency by a government agency, NASA, and then adopted by the private sector. As described by NASA:

> Following the Arab oil embargo of 1973–74, American aviation faced nearly industry-crushing increases in the cost of fuel. In an attempt to make flight operations more efficient, NASA formed the Aircraft Energy Efficiency (ACEE) program. One aspect . . . was handled at Langley Research Center (LaRC). There, a talented engineer named Richard T. Whitcomb elaborated on a concept introduced in the late 1800s . . . wing endplates.
>
> . . . .
>
> On July 24, 1979, the first winglet test flight took off from NASA Dryden Flight Research Center.
>
> . . . .
>
> Over the course of 48 test flights, winglets were proven to reduce wingtip drag, increasing fuel efficiency by 6–7%. Winglets continue to be relied on today by commercial and other airlines for enhanced fuel performance.[28]

---

27. There is an important difference between decisions by aircraft makers to design airplanes—decisions that normally have no governmental energy-efficiency regulation—and decisions by automobile manufacturers to design cars and trucks—decisions that are governed by the CAFE standards. Aircraft makers understand that their customers, the airlines, will make purchase decisions based on full understanding of operating costs of the competing aircraft. Automobile manufacturers understand that their customers are unlikely to calculate the full operating cost of the automobiles they purchase. See discussions of beliefs by automakers about their customers, described in the two National Research Council studies cited in the section on cars and light trucks.

28. "This Month in NASA History: Winglets Helped Save an Industry," NASA's Academy of Program/Project and Engineering Leadership (APPEL), July 22, 2014, appel.nasa.gov/2014/07/22/this-month-in-nasa-history-winglets-helped-save-an-industry/.

Winglets
Source: https://spinoff.nasa.gov/Spinoff2010/t_5.html

Winglets were adopted in early years by Learjet and Gulfstream; Boeing and other companies integrated wingtips into jets, beginning in the 1980s. Winglets have continued to improve since 1979 and have been broadly adopted for medium- and long-haul airplanes.

Another more recent engineering decision was to use new lightweight materials, particularly composites, in airplanes. Both Airbus and Boeing recently introduced large, twin-aisle airplanes—the Boeing 787 Dreamliner and the Airbus A350—with major structures (fuselage, wings, tail) predominantly composites rather than aluminum. The significant weight savings translates directly to fuel consumption savings. Both airplane families provide a 20–25 percent fuel efficiency gain over the models they are designed to replace.

The change in engineering and design was motivated by a desire to reduce fuel costs. As with winglets, smaller manufacturers (Learjet and Gulfstream) had already been incorporating composites, and composites had been used on military fighter jets. Boeing and Airbus were strongly encouraged by their customers, the airlines, to develop these more fuel-efficient airplanes, even though there were alternatives for larger and faster airplanes. Market forces continue to bring enhanced aircraft energy efficiency.

Computing

Ubiquitous in the commercial and industrial sectors of the economy are data centers or other computing clusters. These centers and clusters have become more energy efficient, both by increasing the computing efficiency and by decreasing the energy loss associated with their surroundings, particularly the cooling and other overhead. These changes appear to have been driven by two forces. The first are market forces where there is a recognized need to keep computing and data storage costs under control as the amount of data being used and stored continues to grow exponentially, for example, owners or operators who are either well educated (NREL below) and/or operate on an international level (thus experiencing international forces to keep prices low, similar to that experienced in the aviation industry described above). The

Network Appliance (NetApp®) File Server
Source: http://www.arm.gov/news/facility/post/987

second force has been on the policy front, with significant financial support by the DOE and utilities.

Two press releases, one by the corporation NetApp and the other by the National Renewable Energy Laboratory (NREL), illustrate energy-efficiency gains. Both emphasize the power usage effectiveness (PUE) of their data centers. The PUE is the ratio of total energy used by a computer data facility to energy delivered to computing equipment. If the PUE were 1.0, then all energy used would be delivered to the computing equipment, and if the PUE were 2.0, then the total use of energy would be twice the energy used by the computing equipment itself.

In 2009, NetApp announced that the PUE of their new data center would be 1.2. In comparison to a PUE of 2.0, such a system would reduce energy use by 40 percent. NetApp emphasizes in the press release that the low PUE "will result in NetApp saving $7.3 million a year." Cost reduction motives are consistent with this energy-efficiency initiative by NetApp.

The NREL recently opened its high-performance computing data center. NREL, whose mission includes development of renewable energy technologies and energy-efficiency technologies, announced that its high-performance computing data center would have a PUE of 1.06 or smaller. Such a small PUE is not typical of commercial installations, but it does demonstrate new technology configurations that increase energy efficiency. A demonstration like this provides useful information to private firms, showing both that such small PUEs can be achieved and how one can accomplish the end.

Virtualization of computer servers is a relatively new technology that makes it possible to run multiple operating systems and applications on the same server at the same time. Companies can reduce the number and types of servers for their business applications and thereby reduce the number of computers, the electricity used by those computers, the need for cooling for the computers, and thus the energy used for the cooling system.

It is unlikely that virtualization, such as provided by VMWare, was developed just to improve energy efficiency. However, energy efficiency is part of the reduction in overall costs. And VMWare advertises virtualization as a means of cutting costs for computing facilities. For example, the VMWare website[29] advertises virtualization as "the single most effective way to reduce IT expenses while boosting efficiency and agility—not just for large enterprises, but for small and midsize businesses too." Virtualization enhances energy efficiency, even though energy efficiency has been but a secondary motivation for this computing technology.

For each of the examples discussed above, technology advances, typically motivated by the desire to reduce energy use, have been fundamental to the reductions in energy use. Other changes, however, have not depended as essentially on technology advances, although technology advances did play some role. We turn to several examples of energy efficiency stemming from increased adoption of already-existing technologies.

Changed Adoption of Energy-Efficient Technologies

Some energy-savings technologies, building insulation, insulation on water heaters, programmable thermostats, and variable speed pool pumps, were around well before the 1973–74 energy crisis but have been improved and adopted in greater amounts after the crisis.

Building Insulation

Increased energy prices and increased awareness of energy costs motivated people to install more insulation in homes and other buildings. The standards for insulation were increased for new building construction, with walls and ceilings typically well insulated after the

---

29. http://www.vmware.com/virtualization/overview.html

energy crisis. Existing buildings were more challenging because of the high cost to put insulation into existing walls. But for homes with attics, fiberglass insulating bats could be installed below the attic ceiling. For homes with crawl spaces, but no attics, insulation could be blown into the crawl spaces. Companies were established to provide these installation services, and they actively market their services.

Building Insulation

Source: http://energy.gov/energysaver/articles/adding-insulation-existing-home

The federal government and some state governments, for example, California, offer encouragement to better insulate buildings. For example, on the energy saver website Energy.gov is guidance on the installation of home insulation[30] for those who prefer the do-it-yourself approach rather than contracting with an energy service company.

Other Technologies in Buildings

Insulation is not limited to the building shell. One can wrap a home water heater with a low-cost blanket of insulation, thus reducing energy

---

30. See, for example, energy.gov/energysaver/articles/adding-insulation-existing-home.

cost. Once utilities, state governments, and other entities began publicizing this option, homeowners increased their adoption of this easy, very simple technology.

High energy costs have provided an incentive to install weather stripping and to replace drafty windows. Windows are costly; weather stripping is cheap and can be installed by the homeowner. Weather stripping is now commonly adopted. Energy-efficient installation of new windows is less common.

Programmable thermostats can automatically turn the heating temperature down while families are sleeping, or the cooling temperature up when no one is home. However, there was a problem with the first generation of such thermostats: most people never actually programmed the thermostats.[31] As a result, programmable thermostats had little impact on energy use. But that problem was fixed by makers of the second generation of such thermostats by pre-programming them prior to sale. Thus newly installed thermostats, absent any homeowner programming, automatically turn the heating temperature down at night while families presumably are sleeping and turn the cooling temperature up during the middle of the day when people are generally away from home. The default option for the homeowner is energy-efficient programming. The homeowner can reprogram these thermostats to any desired pattern. But default options are very powerful. The change to pre-programmed thermostats, using an energy-efficient time pattern of heating and cooling, circumvented the problem of non-programmed programmable thermostats.

Some people use plug strips for electronic equipment such as televisions, home entertainment systems, and printers. These plug strips, if turned off systematically, assure that no electricity is consumed for this

---

31. Remember the VCRs that would never be programmed, so they would constantly flash 12:00?

electronic equipment except when the equipment is being actively used. Remotely operated plug strips and those on timers are also available. Although energy management by using plug strips has gained in popularity, it does not appear to have been broadly adopted.

Less well-known is the energy savings from replacing fixed-speed pool pumps with variable-speed pumps for those homes, hotels, or other commercial establishments with swimming pools. In contrast to a traditional fixed-speed pool pump, a variable-speed pool pump can reduce the energy used by 60–90 percent. But to make the change, the pool owner needs to know about this option. Customer-funded utility programs have spent considerable money and effort in the last decade seeking to educate their customers about these savings.

Variable-speed Pool Pump

Source: http://www.hayward-pool.com/shop/en
/pools/maxflo-vs-i-pmmxvs--1

When businesses make a commitment to energy efficiency in their buildings, typically it is not one change, but a large number of changes; each, taken alone, has only a relatively small impact. The results may accumulate slowly but steadily, ultimately leading to large reductions in energy intensity. An example is Walmart. As reported in their *2014 Global Responsibility Report:*

In 2012 and 2013, we completed and/or commenced implementation of numerous energy-efficiency initiatives as we continued to reduce the energy intensity (kWh/sq. ft.) of our facilities worldwide. These measures included installing sales floor LED lighting, high-efficiency (HE) refrigeration units, HE heating, ventilation and air conditioning units, doors on refrigerated cases, parking lot LEDs, energy management systems, voltage optimization systems, LEDs in refrigerated cases, and retro-commissioning of buildings and HE air-handling units.[32]

In the same report, Walmart described the impacts of all the changes taken together:

Toward the end of 2012, we announced that we met our 2005 commitment to r_educe GHG emissions associated with our existing stores, clubs and distribution centers by 20 percent, one year ahead of our seven-year target. The majority of these reductions were driven by energy efficiency.

A few months later, Walmart leaders announced our new goal to reduce the total energy intensity per square foot (kWh/sq. ft.) of all operating facilities by 20 percent by 2020, compared to our 2010 baseline.[33]

In the 2015 report, Walmart described the continuing results:

As of the end of 2014, we were well on our way toward this new goal by operating with 9 percent less energy per square foot compared with our 2010 baseline. This represents a 2 percent improvement since the end of 2013.[34]

32. http://cdn.corporate.walmart.com/db/e1/b551a9db42fd99ea24141f76065f/2014
-global-responsibility-report.pdf

33. Ibid.

34. http://cdn.corporate.walmart.com/f2/b0/5b8e63024998a74b5514e078a4fe/2015
-global-responsibility-report.pdf

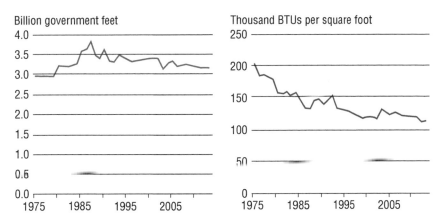

Figure 2.10. Energy Intensity Changes for Federal Government Facilities

Source: Energy Information Administration. http://www.eia.gov/todayinenergy/detail.cfm?id=19851

Efficiency in Federal Government Buildings

Federal government facilities, primarily buildings, have greatly reduced their energy use, through policies and programs implemented by the General Services Administration and others. These measures include installation of efficient lighting, insulation, and building management systems. Data published by the US Energy Information Administration show that the energy intensity of federal government facilities, primarily buildings—measured in terms of thousand BTUs per square foot—has declined by a factor of almost 50 percent from 1975 through 2013.[35] Data are in Figure 2.10.

---

35. According to the Buildings Energy Data Book, in FY 2007, federal buildings accounted for 2.2 percent of all building energy consumption and 0.9 percent of total US energy consumption. Five federal agencies were responsible for 83 percent of all federal building primary energy consumption: the Department of Defense (DOD) (54%), the US Postal Service (USPS) (10%), the Department of Energy (DOE) (10%), the Department of Veterans Affairs

Changed Company Practices

The examples mentioned above have been new/improved technologies and increased adoption of existing technologies. But some changes, particularly in companies, have very little to do with specific technologies but are more related to the behavior of individuals within companies or the organizational behavior of companies.

Reducing Energy Usage as a Profit/Cost Strategy

For companies with high energy costs, deliberate management, designed to enhance energy efficiency, has become an important profit strategy. Many such companies have hired energy coordinators or sustainability coordinators charged with finding ways of reducing energy use so as to reduce costs and increase profits. The functions of energy coordinators vary sharply with the products and processes of the particular company, so generalization is difficult. In what follows are only a few examples.

The breadth of strategies and the payoffs from these strategies can be illustrated by a 2008 press release from DuPont Titanium Technologies, one of the DuPont companies.

The headline proclaimed that energy reduction actions had saved the company $100 million, thus illustrating the financial payoff. The strategies were many but included "tracking down every possible opportunity to save energy. It all adds up, one motor and one valve at a time, day after day." Like much of energy efficiency throughout the economy, industrial energy efficiency is generally not based on a small number of "eureka" moments, but rather an accumulation of many

---

(VA) (6%), and the General Services Administration (GSA) (5%). Note that federal agencies are required to meet energy management mandates outlined by several legal authorities dating back to 1978 (see http://energy.gov/eere/femp/energy-management-mandates-federal-legal -authority).

Dupont's Energy Reductions Save $100M

GreenBiz Staff

Monday, March 3, 2008 - 5:00pm
Since 2001, DuPont Titanium Technologies has saved more than $100 million through cuts in energy consumption achieved through innovation and operational changes.

The company, which is the world's largest titanium dioxide manufacturer, reduced its energy consumption per pound of product by nearly 30 percent.

"There's no one great 'Eureka!' moment in this kind of effort," said Rick Olson, the company's vice president and general manager. "Every one of our sites around the world is tracking down every possible opportunity to save energy. It all adds up, one motor and one valve at a time, day after day. This work is never finished."

The company upgraded the hardware and software in its plant control rooms to give operators more detailed energy information. DuPont also invented new pigment products that can be manufactured using less energy. A new plant DuPont plans to build in China will utilize manufacturing methods that are less energy intensive.

Press Release from DuPont Titanium Technologies

Source: www.greenbiz.com/news/2008/03/03/duponts-energy-reductions-save-100m

actions, each normally having only a very small impact: "one motor and one valve at time." And like much of energy efficiency throughout the economy, gains are based on continuous improvement: "day after day."

And this is not a new phenomenon. For DuPont Titanium Technologies, when the press release was released eight years ago, the changes had been going on for seven years. Finally, the energy-efficiency improvements were not limited to only one class of change, but rather involved innovation of many types: DuPont upgraded hardware and software to give operators more detailed energy information; they invented new products that could be made using less energy; their planned new plant would use less energy.

Data-Driven Industrial Energy Management

Industrial energy management normally needs data on the use of energy in the company's various processes and facilities. And it needs a way of bringing these data to those who can effectively manage the energy use. Microsoft is an example of a company who has done just that, although Microsoft is far from alone.

Microsoft uses extensive data processes for energy-efficiency management. They developed and use software that allows them to track the building performance index of each of their various facilities from a central location (see Figure 2.11, provided courtesy of Microsoft). Energy use is tracked and analyzed at different time scales at the building

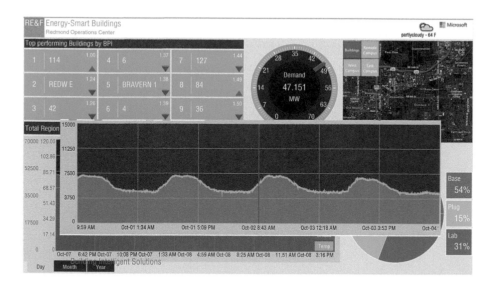

Courtesy of Microsoft.

Figure 2.11. Microsoft Dashboard for Monitoring Energy Use

Source: Microsoft

level and equipment level. According to Microsoft "The team now collects 500 million data transactions every 24 hours, and the smart buildings software presents engineers with prioritized lists of misbehaving equipment. Algorithms can balance out the cost of a fix in terms of money and energy being wasted with other factors such as how much impact fixing it will have on employees who work in that building."[36] Microsoft estimates that this system can reduce energy use by 6–10 percent per year, with capital costs that pay back within one and a half years.[37]

Walmart, a very large user of electricity in its various facilities, has a strong profit motive for managing its energy use. But not all energy use is controlled by Walmart personnel, nor does all energy use follow a predictable pattern. For example, a customer could leave a cooler or freezer door open, thus increasing energy losses. And individual store managers may not be as energy conscious as the corporate standards.

Walmart has built a monitoring and management system to address these challenges. The system relies on a rich array of sensors throughout each of its stores, collecting energy-related information. Data from these sensors are fed back in real time to a monitoring system in Walmart headquarters. Heating and cooling of Walmart stores worldwide is centrally controlled in Bentonville, Arkansas, enabling Walmart to significantly reduce energy consumption.[38] The system includes alarms that alert Walmart personnel when something is amiss. For example, if a freezer or refrigerator door is left open beyond a specified time, the operators in central headquarters are alerted. This allows them to follow a predetermined protocol to alert personnel in the local

---

36. "Now" is 2013, the year the articles were posted (https://www.microsoft.com/en-us/stories/88acres/88-acres-how-microsoft-quietly-built-the-city-of-the-future-chapter-2.aspx).
37. https://www.microsoft.com/en-us/stories/88acres/88-acres-how-microsoft-quietly-built-the-city-of-the-future-chapter-4.aspx
38. http://corporate.walmart.com/_news_/news-archive/2005/01/07/some-facts-about-Walmarts-energy-conservation-measures

store, so the problem can be fixed quickly. The data system allows Walmart to benchmark the various facilities and determine which ones are being managed in a suboptimal fashion. That data-driven understanding then allows Walmart to develop specific interventions for those suboptimal facilities.[39]

Airline Capacity Factor Management

A distant second[40] to cars and trucks for personal mobility is air travel, primarily on commercial airplanes. In addition to the airline technology changes discussed above, airline operations have also focused on fuel savings to reduce costs.[41] The results, measured as reductions in the use of fuel per passenger mile, have been dramatic. Even though air transportation continues to grow rapidly, the amount of jet fuel used has been declining. Consumption of jet fuel in the United States reached a peak in the year 2000 and declined 15 percent from 2000 to 2015. This has been a story of technology, operations, route scheduling, customer incentives, and other yield management. This section focuses primarily on yield management and other changes designed to increase the capacity factor of airlines.

Airlines strive to reach as close as possible to 100 percent seat occupancy, even with passengers making last-minute changes and with weather- and maintenance-related changes in scheduled flights. But near 100 percent occupancy is a daunting challenge.

---

39. Previously in this chapter, I provided some quantification of Walmart's overall energy efficiency progress, only some of which resulted from this data-intense system.

40. The US Department of Transportation estimates that in 2013 there were 4.3 trillion passenger miles on highways (including 0.3 trillion by bus), 0.6 trillion passenger miles by certified air carriers, and 0.04 trillion by rail (http://www.rita.dot.gov/bts/sites/rita.dot.gov.bts /files/publications/national_transportation_statistics/html/table_01_40.html).

41. As noted above, saving fuel has been a continuous objective, as fuel has accounted for 12.5–32.5 percent of total airline costs. Source: J. J. Lee, S. P. Lukachko, I. A. Waitz, and A. Schäfer, "Historical and Future Trends in Aircraft Performance, Cost, and Emissions," *Annual Review of Energy and the Environment* 26 (2001): 167–200.

The number of customers who wish to travel on a particular flight varies systematically with the day of the week, the month, the proximity to holidays, and the state of the economy. These variations can be statistically predicted. But an airline cannot vary the number of airplanes it owns to follow this statistical pattern. It can vary flight schedules based on these fairly predictable periodic changes, but its options are severely constrained by its existing fleet, its access to departure/landing airport gates, its personnel, and the need to have the right aircraft in the right place at the right time.

In addition, there are variations based on weather or mechanical problems that delay flights or cause cancellations. Delays or cancellations imply that passengers with connecting flights may miss those flights and must be accommodated on alternative flights. And a delayed or cancelled flight means that a particular airplane might not be available for its next scheduled flight from what would have been its destination airport. Equipment problems thus can cascade.

The statistical variation in number of people who wish to take a particular flight and the variation in the availability of the flights (based on weather and mechanical issues) makes it virtually impossible to keep 100 percent seat occupancy.

Airlines (and consultants to airlines) have developed complex stochastic, dynamic computer-based optimization processes to manage this random variation. Collectively these processes go under the name of "yield management."[42] Yield management has a goal of maximizing profits from a perishable resource[43] through predicting, monitoring, and influencing the customer decisions. A major part of maximizing

---

42. Yield management is not limited to airlines, but is widely practiced in other service industries with perishable commodities, for example, hotel rooms and telecommunications.

43. An airline seat is a perishable resource because once the airplane has taken off, the unfilled seats cannot be sold.

profits[44] is maintaining near 100 percent seat occupancy; higher occupancy rates reduce the energy used per passenger mile, without greatly influencing the energy used per seat mile.

Yield management processes include variable pricing for a given route, based on statistical analysis of the number of passengers to be expected and on updating of the expectations based on actual reservations. The pricing for a given flight can change from day to day, based on this updated information. Pricing involves market segmentation, selling travel with different restrictions at different prices (for example, length of stay requirements, cancellation fees) as well as deliberate overbooking, a practice that frustrates travelers but helps to increase the fraction of seats occupied. Yield management processes also involve incentives for passengers not to change their scheduled flights until the last minute, at which time the airline has very good information about whether a given flight will be full. For example, many airlines charge large fees for changing the scheduled flight more than a day ahead but will allow changes at no cost to the passenger within 24 hours of the flight.

The resultant reductions in cost of travel have for the most part led to reductions in the cost of air travel in the United States, an overall economic benefit of these private-sector efficiency gains.[45]

The combination of the technology changes discussed in a previous section and the changes in airline practices have greatly reduced the fuel use per passenger mile. These changes are shown in Figure 2.12,

---

44. There are other elements of yield management that increase profits but have no impact on energy efficiency. And other strategies have increased profits of airlines. For example, mergers of airlines have resulted in reduced numbers of competing flights, allowed fare increases, and increased the capacity factor. I have not analyzed whether these changes have led to overall economic benefits. For a more complete discussion, see *Aviation Industry Performance: A Review of the Aviation Industry, 2008–2011*. Number: CC-2012-029, September 24, 2012, Department of Transportation, Office of Inspector General, https://www.oig.dot.gov/sites/default/files /Aviation Industry Performance%5E9-24-12.pdf.

45. There are also small economic gains based on the downward pressure on the price of imported oil, as discussed above in reference to light-duty-vehicle fuel economy gains.

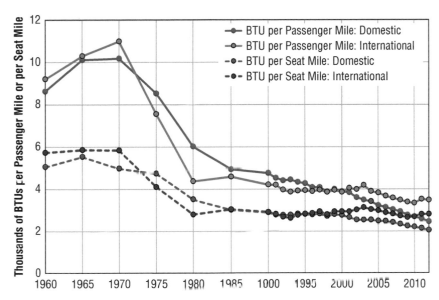

Figure 2.12. Energy Intensity of Certificated Air Carriers

Source: US Department of Transportation, Bureau of Transportation Statistics, http://www.rita.dot.gov
/bts/sites/rita.dot.gov.bts/files/publications/national_transportation_statistics/html/table_04_21.html

which plots the energy intensity—the fuel used per passenger mile—
of certified air carriers[46] in the United States from 1960 through 2012.
Fuel use per passenger mile is shown by the solid red and blue lines. In
1976, air carriers used over 10,000 BTUs per passenger mile for both
domestic and international operations. By 2012, fuel use for domestic
operations had dropped to under 3000 BTUs per passenger mile and
for international operations to under 4000.

Figure 2.12 also shows a second measure of energy intensity—the
fuel used per seat mile. Fuel use per seat mile is shown by the broken dark
red and dark blue lines. Fuel used per passenger mile has declined more

---

46. Includes all commercial airlines flying passengers and/or cargo. Bureau of Transportation
Statistics, US DOT, http://www.rita.dot.gov/bts/sites/rita.dot.gov.bts/files/table_04_22_1.xlsx.

rapidly than fuel used per seat mile. These two measures of intensity differ primarily as a result of successful airline efforts to increase the capacity factor, the fraction of seats that are full.

As shown in Figure 2.12, fuel used per passenger mile by 2012 declined to 29 percent and 46 percent of 1975 levels for domestic and international operations, respectively. For better understanding, fuel used per passenger mile can be decomposed into two factors multiplied by each other: fuel use per seat mile and seat miles per passenger mile. (Mathematically, seat miles per passenger mile is the inverse of capacity factor.) By 2012, fuel used per seat mile had declined to 43 percent and 68 percent of 1975 levels for domestic and international operations, respectively. Seat miles per passenger mile decreased to 67 percent of the 1975 level for both international and domestic operations.

For international operations, the change in fuel used per seat mile, resulting from a combination of technology and practice changes, and the change in capacity factor, primarily resulting from airline practice changes, have had similar impacts on energy per passenger mile. For domestic operations, the change in fuel used per seat mile has been the largest factor.

The dramatic reduction in energy intensity of airline operations has implied that even though air transportation continues to grow rapidly, the amount of jet fuel used has been declining. These data[47] are shown in Figure 2.13. Passenger miles are shown by the broken red and blue lines. Fuel consumption is shown by the solid bright red and dark blue lines.

Airlines, with fuel costs as a large fraction of their operating costs, have been highly motivated to find ways of decreasing the use of fuel per passenger mile. Less fuel per passenger mile directly implies less cost per passenger. An airline either can keep all the cost savings as an increase in profit or can pass some of the cost savings to customers, as

---

47. http://www.rita.dot.gov/bts/sites/rita.dot.gov.bts/files/table_04_22_1.xlsx

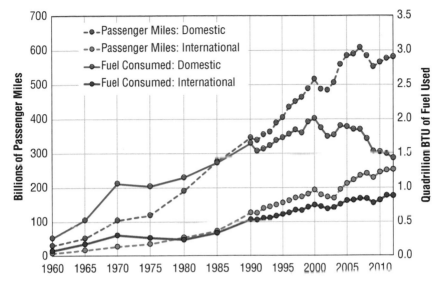

Figure 2.13. Airline Passenger Miles and Fuel Use

Source: http://www.rita.dot.gov/bts/sites/rita.dot.gov.bts/files/table_04_22_1.xlsx

lower fares. Fares need not be reduced uniformly, but can be selectively reduced within a yield management system. Lower fares themselves tend to increase the number of passengers, thereby increasing profit. Lower fares selectively applied through a yield management system can move the airline closer to 100 percent seat occupancy. Whichever strategy an airline uses, if it can reduce fuel use per passenger mile, it can increase its profits. And if its competitors reduce fuel use per passenger mile, and it does not, then it loses customers and therefore profit.

In addition, airline mergers have enabled the combined airlines to reduce previously competing flights and redundant hub operations. They have cut back on available capacity by reducing the number of flights.[48]

---

48. See *Aviation Industry Performance: A Review of the Aviation Industry, 2008–2011*. Number: CC-2012-029, Department of Transportation, Office of Inspector General, September 24, 2012, https://www.oig.dot.gov/sites/default/files/Aviation Industry Performance%5E9-24-12.pdf.

These changes have worked along with yield management systems to increase the capacity factor of airlines, further reducing the fuel use per passenger mile without greatly influencing the fuel use per seat mile.[49]

This strong private sector incentive for reducing fuel use per passenger mile has led to the remarkable energy-efficiency gains in airlines, through a combination of factors. Jet engine makers—such as General Electric, Rolls-Royce Holdings, and Pratt & Whitney—have financial incentives to design engines that are more fuel efficient as they compete with each other. Aircraft manufacturers—such as Boeing, Airbus, Bombardier—have financial incentives to design their airplanes to use less fuel per seat mile, both through design of the airplane body and through incorporation of energy-efficient engines, as they compete with each other for sales to airlines. Airlines—such as United, Delta, American, Southwest—have financial incentives to buy fuel-efficient aircraft and to operate them in an efficient manner. Collectively, the result has been sharp decreases in fuel use per seat mile. In addition, they have incentives to keep as close as possible to 100 percent seat occupancy, thereby further reducing fuel use per passenger mile. The net result, shown in Figure 2.12 and Figure 2.13, has been a dramatic increase in energy efficiency of airline travel.

Behavioral Strategies

Other companies use behavioral strategies to empower employees to find energy-efficiency improvements. For example, in Raytheon Corporation employees can become "Energy Champions" or "Energy Citizens." Raytheon describes the roles as follows:

---

49. I have not tried to determine whether these reductions in the number of competing airlines and the number of flights have been economically efficient or to determine whether they represent exercise of airline market power.

Energy Champions comprise a volunteer network of employees who develop energy conservation strategies and measures for their work areas, and ensure savings are maintained. These role models foster a culture of energy conservation and efficiency among their team members through sharing progress, lessons learned, and innovative practices.

. . .

To qualify as an Energy Citizen, employees take an annual online quiz, which assesses their conservation practices at work and home. The questionnaire also serves to further educate employees through information hotlinks. Employees who pass the quiz receive a lapel pin and a note from top management to commend their efforts. During 2009, 40 percent (over 28,000) of employees qualified.[50]

Raytheon reduced energy consumption by 15 percent between 2008 and 2014, resulting in a $55 million cumulative energy cost savings between those years.[51] How much of that has been the result of the behavioral strategies is not clear. What is clear is that Raytheon has been reducing its energy use by about 2.5 percent per year.

For Dow Chemical Company, energy is a large fraction of their overall costs. After their Board of Directors and CEO set an energy reduction goal for the entire corporation, meeting the goal required energy reductions for all of their activities. Many of the strategies they adopted were behavioral. For example, formal reports by division managers required an energy element describing actions taken in that division. Because the element was required, the managers had to pay attention to finding ways of improving energy efficiency. Energy-specific suggestion boxes were installed throughout their facilities. All employees learned that wasting energy increased costs, and were

50. http://www.mass.gov/eea/docs/doer/energy-efficiency/raytheon.pdf

51. http://www.raytheon.com/responsibility/rtnwcm/groups/public/documents/content/crr_pdf.pdf

rewarded for bringing energy waste to the attention of their managers so the problem could be fixed. By primarily keeping energy at the front of everyone's mind, since 1990 Dow Chemical has been successful in reducing energy intensity, measured in BTUs per pound of product, by more than 40 percent. Estimates by Dow Chemical are that this has saved the company more than $24 billion.[52]

Commercial Building Retrofits

A changed practice is to retrofit older commercial buildings in order to reduce energy costs. With increased energy prices, such energy-efficiency retrofits have become profitable for many older commercial buildings.

The Empire State Building is a particularly visible and striking example of a successful energy-efficiency retrofit. In 2009, the Empire State Realty Trust initiated a $13.2 million retrofit to "reduce costs, increase real estate value and protect the environment."[53] The retrofit included changes to the base building energy system (e.g., replacing the chiller for building cooling) and many changes within tenant spaces (e.g., building windows and lighting systems).

This project suggests that retrofits of older commercial buildings can be very profitable.[54] As such, it is likely to encourage others to examine energy-efficiency retrofits on other buildings.

---

52. http://www.dow.com/en-us/science-and-sustainability/sustainability-reporting/energy-efficiency-conservation

53. http://www.esbnyc.com/sites/default/files/esb_year_three_press_release_final.pdf

54. The retrofit did receive capital subsidies from the state of New York, but, according to data from the Empire State Realty Trust, the project would have been profitable even without the subsidies. According to the Empire State Realty Trust, energy savings have exceeded the projections made at the time of the retrofit. "Over the past three years, the program has generated a total of approximately $7.5 million in energy savings" according to an August 2014 press release. "The savings is the result of continued enhancement of the iconic building's new systems and the addition of many new tenants occupying hundreds of thousands of square feet of office space retrofitted according to program guidelines." See http://www.esbnyc.com/sites/default/files/esb_year_three_press_release_final.pdf.

Particularly interesting are the assessment of an array of projects within the retrofit, the selection of a group of cost-effective projects, the unusual practice of communicating publicly their processes and their choices, and their description of the invisibility of the energy-efficiency enhancements.

According to the Empire State Realty Trust:

> The recommended package of 8 projects saves nearly 40% of the energy at the Empire State Building, but almost all the changes are unnoticeable from the outside and by visitors to the 86th floor Observatory.
>
> The sustainable transformation of the Empire State Building has happened inconspicuously—and the huge energy savings come from the interaction between the multiple retrofits.[55]

As noted above, the retrofit was not one single project, but a group of eight major changes. Each change taken alone had relatively little impact on the energy use in the building: each reduced energy use between 2 percent and 9 percent, as shown[56] in Figure 2.14. But the collection of all projects together reduced electricity use by at least 38 percent.

For the Empire State Building, as is the case throughout the United States, the energy-efficiency enhancements were the result of many relatively small changes, most invisible to outside observers but accumulating to large reductions in energy use.

Contracts/Collaborations to Overcome Split Incentive Problems

Chapter 1 identified that split incentive problems get in the way of full implementation of energy-efficiency options (see Table 1.1. Some Barriers

---

55. http://www.esbnyc.com/esb-sustainability/project

56. Empire State Realty Trust Powerpoint: "Energy Efficiency in the Built Environment: Learnings from Ground-breaking Work at the Empire State Building," courtesy of Tony Malkin.

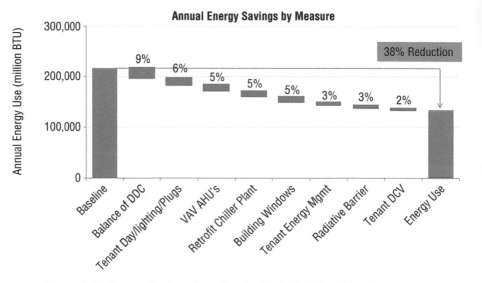

Figure 2.14. Energy Savings from Empire State Building Retrofit

Source: Tony Malkin

to Energy-Use Optimality). Briefly stated, in office buildings, the landlord is responsible for the upkeep of the physical characteristics of the building and the common areas, and the tenants are responsible for energy costs within their spaces. Both optimize for their own economic interests, and as a result, many energy-efficiency improvements that could be mutually beneficial are not undertaken.[57]

A changed practice, designed to address this split incentive problem, is the negotiation of performance contracts between the landlord and the tenants for investing in mutually beneficial energy-efficiency improvements. Energy decisions within tenant spaces provide a large

57. Among other places, these issues are discussed in Marilyn A. Brown, "Market Failures and Barriers as a Basis for Clean Energy Policies," *Energy Policy* 29 (2001): 1197–1207. Or see K. Gillingham and J. Sweeney, "Barriers to Implementing Low Carbon Technologies," *Climate Change Economics* 3, no. 4 (2012): 1–25. doi: 10.1142/S2010007812500194.

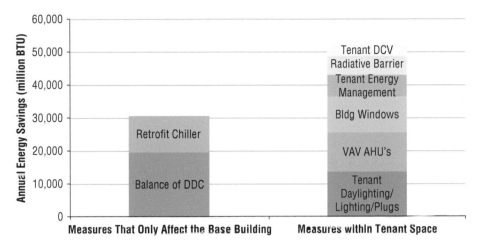

Figure 2.15. Base Building vs. Tenant Space Improvements: ESB

Source: Tony Malkin

fraction of the opportunities for energy efficiency, and it is within these tenant spaces that performance contracts can be particularly useful.

Such performance contracts between landlord and tenant can be illustrated by the Empire State Building (ESB) energy-efficiency retrofit discussed above. For the ESB retrofit project, Figure 2.15 shows the energy savings for the base building versus the savings within the tenant space.[58] For this retrofit, the majority of the savings were within the tenant space.

The building owner, the Empire State Realty Trust (ESRT), negotiated the bulk of the energy-efficiency upgrades in tenant spaces as a shared responsibility under performance contracts between the building owner and the tenants. ("Five of the eight projects will be

---

58. Empire State Realty Trust Powerpoint: "Energy Efficiency in the Built Environment."

implemented using a performance contract. . . . These projects will save $2.4 million of the total $4.4 million annual savings.")[59]

But creating performance contracts required trust: trust built on collaboration between the building owner and the tenants, long-term leases, a joint understanding of the benefits of efficiency investments, and a mechanism for multiple tenants to work with the landlord in a coordinated fashion. In the case of the Empire State Building, a collaborative process between the building owner, the ESRT, and industry partners—the Tenant Energy Analysis and Metrics (TEAM)[60]— provided the framework for the collaboration.

A documented example of the TEAM process is the experience of Coty, a leader in beauty products, which leased four additional floors of the ESB in 2012. In designing and constructing the space Coty had three major goals: "to increase energy efficiency, reduce costs, and ensure the best possible environment for its employees."[61] In partnership with ESRT and the TEAM partners, Coty evaluated an integrated package of energy performance measures for the first two floors. The process included energy modeling and incremental costing information, based on estimated impacts and subsequently on measured performance. The outputs of the value analysis model are shown in Table 2.2.[62]

The chosen measures were incorporated into the space design and were used to inform the next stage. Coty estimates that over the term of the 17-year lease, the energy-efficiency investments will save about $486,000 and cost about $113,000 (discounted present values.)

59. http://www.esbnyc.com/esb-sustainability/project
60. Additional information and resources for the TEAM process can be found at http://www .teamforoffices.com.
61. http://teamforoffices.com/case-study-coty-inc. Thanks to Coty, Wendy Fok, and Tony Malkin for providing the information above.
62. Ibid.

EPM DESCRIPTION	ORIGINAL MODEL			CALIBRATED MODEL		
	ANNUAL TENANT ELECTRICITY SAVINGS (KWH)	PERCENT SAVINGS	ANNUAL COST SAVINGS	ANNUAL TENANT ELECTRICITY SAVINGS (KWH)	PERCENT SAVINGS	ANNUAL COST SAVINGS
B-2 ASHRAE-2007 Baseline	n/a	n/a	n/a	n/a	n/a	n/a
L-1 High Efficiency Lighting Design	91,574 kWh	17.0%	$15,110	113,011 kWh	14.3%	$18,647
L-2 Daylight Harvesting	37,695 kWh	7.0%	$6,220	27,641 kWh	3.5%	$4,561
L-3 High Efficiency VAV HVAC	34,428 kWh	6.4%	$5,681	62,462 kWh	7.9%	$10,306
L-4 Demand Controlled Ventilation	48 kWh	0.0%	$8	−279 kWh	0.0%	($46)
P-1 Eliminate Noise Traps	2,133 kWh	0.4%	$352	4,142 kWh	0.5%	$683

Table 2.2. Coty Energy Model Output by Measure

EPM DESCRIPTION		ORIGINAL MODEL			CALIBRATED MODEL		
		ANNUAL TENANT ELECTRICITY SAVINGS (KWH)	PERCENT SAVINGS	ANNUAL COST SAVINGS	ANNUAL TENANT ELECTRICITY SAVINGS (KWH)	PERCENT SAVINGS	ANNUAL COST SAVINGS
P-2	Plug Load Management - Energy Star Equipment	20,953 kWh	3.9%	$3,457	36,472 kWh	4.6%	$6,018
X-1	Plug Load Management - Occupancy Sensor Strips	14,596 kWh	2.7%	$3,457	25,254 kWh	3.2%	$4,167
X-2	Plug Load Management - NightWatchman Software	19,401 kWh	3.6%	$3,457	41,482 kWh	5.2%	$6,845
X-3	Low Velocity Air Handlers	9,795 kWh	1.8%	$3,457	18,958 kWh	2.4%	$3,128
	Implemented Package Total (L-1 to P-2)	*186,832 kWh*	*34.7%*	*$3,457*	*243,449 kWh*	*30.7%*	*$40,169*

Table 2.2. (continued)

PHASE 1 BUILD-OUT (FLOORS 16 AND 17)	PROJECT DESIGN		M&V CALIBRATION	
Square Footage	80,000 square feet		80,000 square feet	
Modeled Energy Reduction	32%		30.7%	
Annual Electricity Reduction	186,396 kWh	1.2 kWh/SF	243,449 kWh	1.5 kWh/SF
Total Electricity Savings over Lease Term	3.2 GWh	19.9 kWh/SF	4.1 GWh	26.0 kWh/SF
Incremental Implementation Cost	$144,200	$0.91/SF	$144,200	$0.91/SF
Energy Modeling Soft Cost	$9,000	$0.25/SF	$9,000	$0.25/SF
State Incentives	$39,582	$0.06/SF	$39,582	$0.06/SF
Adjusted Incremental Implementation Cost	$113,618	$0.71/SF	$113,618	$0.71/SF
Total Electricity Costs Savings over Lease Term	$548,317	$3.44/SF	$716,148	$4.49/SF
Electricity Cost Savings over Lease Term (Present Value)	$371,893	$2.33/SF	$485,723	$3.05/SF
Net Present Value of Project Investment	$258,275	$1.62/SF	$372,105	$2.34/SF
Return on Investment over Lease Term	227%		328%	
Internal Rate of Return	32.1%		44%	
Payback Period (with incentives)	3.5 years		2.7 years	

Table 2.3. Coty Project Information and Projected and Measured Performance

The projected design performance and the measured performance to date are summarized in Table 2.3.

Coty and ESRT have made these results public so others can understand how to adopt the process for their own decisions.[63]

---

63. See http://teamforoffices.com/case-study-coty-inc.

Incentives: Internal Carbon Pricing

Another changed practice involves internal pricing systems that systematically influence the energy-use decisions made throughout a company. Such systems, if fully implemented within a company, create similar incentives to external energy price increases. But, as opposed to external energy price increases, the company does not have to pay the costs that would have been associated with the increased external cost.

One broadly adopted internal price system—an internal carbon price—is not strictly for energy but also for carbon dioxide emissions. But for most companies, the primary release of carbon dioxide emissions is through use of energy. Thus an internal price on carbon is almost equivalent to an internal tax on energy. For many companies, this carbon tax is related to the perceived risks from future regulations or carbon pricing.

An internal carbon price provides corporation-wide incentives for reducing energy use. Perhaps as importantly, the pricing brings energy and carbon costs to the attention of the managers at all levels of the corporation.

Figure 2.16 lists the almost one hundred North American companies that have adopted an internal price on carbon as of 2015, according to the Carbon Disclosure Project, a not-for-profit organization that promotes voluntary carbon pricing.[64]

Quoting from the Carbon Disclosure Project "Putting a Price on Risk: Carbon Pricing in the Corporate World":

> A growing number of U.S. and Canadian companies (more than doubling from 2014 to a total of 97 in 2015) are assigning an internal price to their carbon emissions. These include highly trusted consumer brands such as Colgate-Palmolive and Campbell's Soup, global industrials

---

64. http://www.cdp.net/CDPResults/carbon-pricing-in-the-corporate-world.pdf, pp. 61, 62.

Companies currently using an internal carbon price

	Company	Country	Price (US$)
Consumer Discretion-ary	ARGENT ASSOCIATES INC	USA	
	Baccus Global LLC	USA	
	Canadian Tire Corporation, Limited	Canada	6.36–30
	Fruit of the Loom	USA	
	General Motors Company	USA	5
	Walt Disney Company	USA	10–20
Consumer Staples	Archer Daniels Midland	USA	
	Campbell Soup Company	USA	
	Chicken of the Sea Intl	USA	10.25
	Colgate Palmolive Company	USA	
	Dean Foods Company	USA	
	Hormel Foods	USA	
	Pacific Coast Producers	USA	
	WhiteWave Foods	USA	
Energy	Apache Corporation	USA	
	ARC Resources Ltd.	Canada	3.77–22.60
	Canadian Oil Sands Limited	Canada	11.3
	Cenovus Energy Inc.	Canada	11.30–48.96
	Chevron Corporation	USA	
	ConocoPhillips	USA	6.0–51.0
	Enbridge Inc.	Canada	150.66
	Encana Corporation	Canada	15.07–94.16
	Exxon Mobil Corporation	USA	80
	Hess Corporation	USA	
	Husky Energy Inc.	Canada	
	Imperial Oil	Canada	80
	Keyera Corp.	Canada	
	Occidental Petroleum Corporation	USA	
	Pengrowth Energy Corporation	Canada	
	Suncor Energy Inc.	Canada	11.30–41.43
	TransCanada Corporation	Canada	
	Vermilion Energy Inc.	Canada	11.30–24.69
Financials	Bank of Montreal	Canada	
	BNY Mellon	USA	23.87
	Goldman Sachs Group Inc.	USA	
	TD Bank Group	Canada	7.53
	Wells Fargo & Company	USA	
Health Care	Allergan, Inc.	USA	
Industrials	Covanta Energy Corporation	USA	
	Cummins Inc.	USA	
	Delta Air Lines	USA	

Figure 2.16. Internal Carbon Prices: North American Corporations

Source: www.cdp.net/CDPResults/carbon-pricing-in-the-corporate-world.pdf, pp. 61, 62

	Company	Country	Price (US$)
Industrials, continued	General Electric Company	USA	
	Owens Corning	USA	10.0–60.0
	Parker-Hannifin Corporation	USA	
	Stanley Black & Decker, Inc.	USA	18.0–150.0
	Tennant Company	USA	
Information Technology	Adobe Systems, Inc.	USA	
	ASOCIAR LLC	USA	
	Google Inc.	USA	14
	Microsoft Corporation	USA	4.4
	PMC-Sierra, Inc.	USA	
Materials	Agrium Inc.	Canada	11.3
	Barrick Gold Corporation	Canada	24.15
	Caraustar Industries, Inc.	USA	
	Catalyst Paper Corporation	Canada	22.6
	E.I. du Pont de Nemours and Company	USA	
	Eastman Chemical Company	USA	
	Hammond	USA	
	HudBay Minerals Inc.	Canada	15.07-37.66
	PaperWorks Industries Inc	USA	
	Resolute Forest Products Inc.	Canada	
	Teck Resources Limited	Canada	11.30–30.13
	The Dow Chemical Company	USA	
Telecom. Services	Genband	USA	
	World Wide Technology Holding Company	USA	
Utilities	Ameren Corporation	USA	23–53
	American Electric Power Company, Inc.	USA	
	Capital Power Corporation	Canada	
	CMS Energy Corporation	USA	
	Consolidated Edison, Inc.	USA	
	DTE Energy Company	USA	
	Duke Energy Corporation	USA	
	Entergy Corporation	USA	
	Eversource Energy	USA	
	Exelon Corporation	USA	
	Idacorp Inc	USA	
	Los Angeles Department of Water and Power	USA	12.45–35.90
	NiSource Inc.	USA	20
	NRG Energy Inc	USA	
	OGE Energy Corp.	USA	
	Pinnacle West Capital Corporation	USA	
	Sempra Energy	USA	13.06
	TransAlta Corporation	Canada	11.30–22.60
	Xcel Energy Inc.	USA	9.0–34.0

Figure 2.16. (continued)

such as General Motors, and financial giants such as top-ten asset size ranked TD Bank. Global companies are voluntarily enacting pricing despite the patchwork of state based regulations, partly as a way of addressing mandatory carbon pricing to which they may be subjected via regulatory regimes in other regions.[65]

In Summary

Since the energy crisis in 1973, improved technologies, changed implementation of technologies and changed company or individual practices have been broadly distributed throughout the US economy. With the exception of CAFE standards for light-duty vehicles, most of the energy-efficiency improvements described above were implemented at a building, company, or family level. Taken alone, each building-level, company-level, or family-level change had relatively little impact on overall national energy use. Many of the larger-scale changes within a company or building were the result of a series of small changes (e.g., changing out lightbulbs, replacing a motor or a valve), which individually have been rather mundane. Most were invisible to outside observers and often even to insiders within an organization.

However, because so many businesses and individuals across the US economy have implemented energy-efficient practices or technologies, and have done so in many different ways, the cumulative, collective impact has had large industry-wide and economy-wide impacts. These cumulative impacts are quantified in the next chapter.

The changes described in this chapter were chosen as important examples of some specific energy-efficiency enhancements, with no attempt to be comprehensive. There are many other changes, too numerous to discuss here. This chapter has described a pattern of large percentage reductions in energy use for specific technologies—cars,

---

65. Ibid., p. 4.

trucks, airplanes, lights, refrigeration. For more integrated systems, such as homes, commercial buildings, and factories, generalization is more difficult, except for federal government buildings. Large reductions have occurred for some buildings and companies, and little changes for others. However, as is shown in the next chapter, overall the impact has been large; an accumulation of many small changes, broadly distributed, has had a large impact on energy use.

In order to examine the overall impact of all of these changes, the next chapter presents aggregate statistics of energy use in the United States, examines changes in trends of energy use after the energy crisis of 1973–74, and discusses the role of energy efficiency on those aggregate trends.

3

Energy Efficiency and Aggregate Energy Intensity in the United States—1950 through 2014

Aggregate data, as summarized below, shows the changes in the energy intensity of the economy or, equivalently, changes in energy productivity. Some reductions in energy intensity have been the result of energy-efficiency enhancements of the types described in Chapter 2. But some have resulted from the continuation of historical trends of economic change, such as were occurring prior to the energy crisis of 1973–74.

In order to estimate how much of the energy-intensity changes have been the result of increases in the rate of energy efficiency improvement from the pre-1973 trends, I modeled a "limited-energy-efficiency" benchmark, a representation of what would have happened without enhanced energy-efficiency gains in the economy subsequent to the 1973–74 energy crisis. The limited-energy-efficiency benchmark is constructed from the time before energy was broadly seen as an important issue, the time before the energy crisis in 1973–74. Using this benchmark, I examine three different times: the pre-energy-crisis period, the time during and immediately following the energy crisis, and the subsequent years.

The Pre-Energy-Crisis Period: 1950 to 1973

After World War II, the United States paid very little attention to the use of energy until the energy crisis of 1973–74. Figure 3.1, using publicly available data provided by the US Energy Information Administration, shows the evolution of primary energy supply and use between

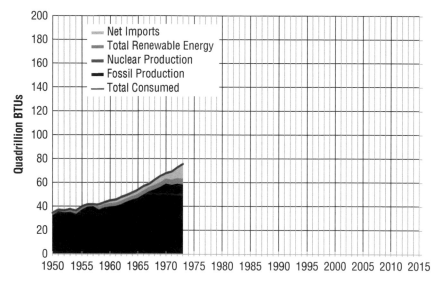

Figure 3.1. US Energy Production and Use to the Time of the 1973 Oil Crisis

Source: EIA, Monthly Energy Review

1950 and 1973. The units of energy production and use are quadrillion BTUs, or, in shorthand, "quads" of primary energy.[1]

Figure 3.1 shows that the use of energy grew almost as rapidly as GDP. This result is also shown in Figure 3.2 by plotting the energy intensity of the US economy—measured in terms of thousands of BTUs per 2009 dollars of GDP—from 1950 through 1973.

Figure 3.1 also shows that the domestic production of energy was based primarily upon fossil fuels and secondarily on renewable forms of energy (hydropower and wood), with a tiny production of nuclear

---

1. Data in this report are of primary energy consumption or production, all measured in terms of quadrillion BTUs of energy (10^{15} BTUs) also referred to as "quads" of energy per year. A quad is 1.055 exajoules of energy (10^{18} joules). Over the course of a year, 0.456 million barrels per day of crude oil imports have primary energy content of 1 quad.

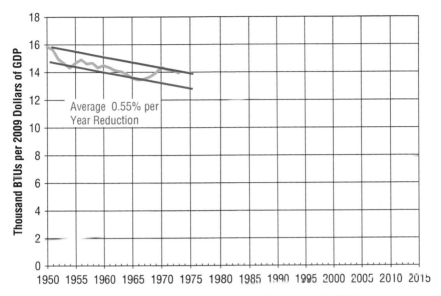

Figure 3.2. Energy Intensity of US Economy, 1950 through 1973

power. It shows net energy imports (in gray) growing rapidly, roughly tripling between 1970 and 1973.

Figure 3.2 includes the actual energy intensity of the economy in blue. The two[2] downward-sloping red lines show the average rate of energy-intensity reduction during this 23-year period. Both red lines show an average annual reduction of energy intensity of 0.55 percent per year.

An extrapolation of trends from the 1950–73 period, a time when little attention was paid to energy use and energy efficiency, allows estimation of what would have happened in the US economy without the

---

2. The lower red line closely approximates the actual energy intensity for most of the years. The upper red line connects actual energy intensity in 1950 to actual energy intensity in 1973. Either way of approximating the changes gives a decline rate of 0.55 percent per year.

enhanced energy-efficiency gains initiated by the energy crisis of 1973–74.

This extrapolation is called the "limited-energy-efficiency" benchmark, although more precisely it might be called the "pre-energy-crisis-historical-rate-of-energy-intensity-reductions" benchmark, a nomenclature that would be too clumsy for this book and will not be used again.

Absent the increased rate of energy-efficiency improvements after the energy crisis, I estimate that energy intensity would have decreased by 0.55 percent per year in the limited-energy-efficiency benchmark.[3] That is, in the "limited-energy-efficiency" benchmark, energy use subsequent to 1973 would likely have grown along with the economy minus 0.55 percent per year.

The "limited-energy-efficiency" benchmark is shown in Figure 3.3. This figure shows that had there been only limited energy-efficiency gains from 1973 through 2014, consistent with the 1950–73 experience, energy use would have continued to grow only somewhat less rapidly than the growth in US GDP, reaching about 180 quadrillion BTUs by the year 2014.

The Energy Crisis: 1973–74

In 1973, the complacent world of energy was fundamentally altered. In October of that year, Egypt and Syria launched an attack against Israel, beginning the short Yom Kippur War. After the United States supplied Israel with arms, Arab members of OPEC—Organization of Petroleum Exporting Countries—initiated an oil embargo against several

---

3. Note that using the limited-energy-efficiency benchmark rather than a benchmark of absolutely no energy-efficiency improvements leads to an underestimate of the impact of energy efficiency on energy use. Thus subsequent estimates of the impacts of energy efficiency should be seen as underestimates of energy-efficiency impacts.

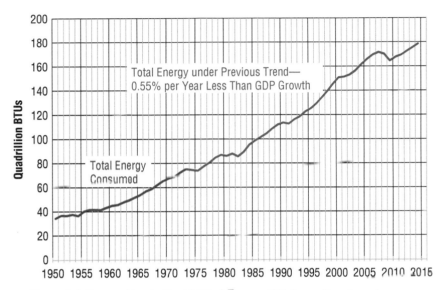

Figure 3.3. Energy Use in the Limited-Energy-Efficiency Benchmark

countries—Canada, Japan, the Netherlands, the United Kingdom and the United States. The embargo itself could be circumvented because oil tankers could be rerouted at sea, and international trade patterns of oil could be shifted. However, along with the embargo, Arab OPEC members reduced their production and exports of oil. As a result, there was a shortage of oil in the world market. Quickly thereafter, the world price of oil tripled, increasing in nominal dollars from about $4 per barrel to almost $12. Adjusted for inflation, the oil price in 2015 dollars increased from about $20 to $60. Data on the world oil price, measured in terms of the price of crude oil imported into the United States, are shown in Figure 3.4.

The Yom Kippur War lasted only about three weeks, but the oil export reductions continued for years afterward. OPEC had discovered that it could exercise market power, pushing up crude oil prices by

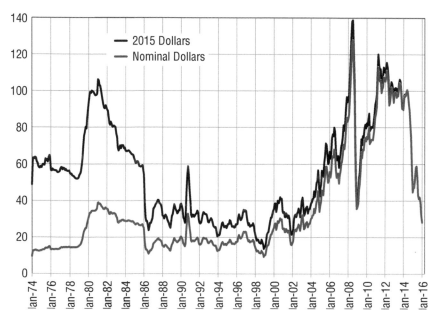

Figure 3.4. Crude Oil Nominal Prices and Real Prices, 2015 Dollars

Source: Energy Information Administration

reducing world supply of oil. As a consequence, the oil price remained high for almost a decade before crashing in 1986.[4]

The embargo, coming when US oil imports were rapidly increasing, coupled with the sudden increase in the oil price, had severe consequences in the United States. This was a period of high inflation and the United States had imposed wage and price controls. These controls included products produced from oil, such as gasoline, so the gasoline

---

4. Until 1986, OPEC nations had agreed on the high oil prices, and Saudi Arabia had kept OPEC power by serving as a swing producer of oil. However, by 1986, decreases in world oil demand and increases in world production capacity had reduced Saudi Arabia sales from about 10 million barrels per day to about 3.5 million barrels per day, and sales were continuing to decrease. Selling very little oil, even at a high price, had put Saudi Arabia in a difficult economic situation. In 1986, Saudi Arabia abandoned its role as swing producer and began producing at full capacity, putting the world price of oil into free fall.

shortages associated with the embargo and the reduced worldwide supply of oil were not eliminated by retail price increases. There were gasoline lines, often hours long, and stations closed on weekends as gasoline stations sold out their limited allotment of fuel, often within hours. Uneconomical energy conservation was imposed on those who could not readily purchase gasoline.

During the energy crisis, in late 1973, President Nixon announced "Project Independence," with a goal of achieving US energy self sufficiency by 1985. Energy policy, and particularly energy security, had become central to US policy initiatives. In support of the policy goal, in 1973 the Nixon Administration created several energy-related offices,[5] which subsequently were combined into the Federal Energy Administration (FEA) in 1974. The FEA was responsible for fuel allocation, pricing regulation, energy data and analysis, energy supply expansion, and planning for energy efficiency. The Energy Research and Development Administration (ERDA), created by the Energy Reorganization Act of 1974, was charged with managing the energy research and development, nuclear weapons, and naval reactors programs.[6]

Also in 1974 the International Energy Agency (IEA) was founded as a Paris-based intergovernmental organization under the Organisation for Economic Co-operation and Development (OECD).

These organizations became governmental foci of energy policy making, energy research, and energy information, continuing to bring attention to policy initiatives and regulations, and enhancing energy awareness well after the crisis conditions were over.

---

5. Within the Department of the Interior were several offices—the Office of Energy Conservation, the Office of Energy Data and Analysis, and the Office of Petroleum Allocation—and in the Executive Office of the President was the National Energy Office and the Federal Energy Policy Office. The FEA was established by the Federal Energy Administration Act of 1974, absorbing the Federal Energy Office and the Offices of Oil and Gas, Energy Conservation, Energy Data and Analysis, and Petroleum Allocation.

6. Subsequently, in 1977, these two agencies became part of the newly created Department of Energy (DOE).

After President Nixon resigned, the Ford administration continued developing Project Independence. The *Project Independence Report*, published by the US Federal Energy Administration in 1974, identified four broad strategic alternatives for US national energy policies to achieve US energy self-sufficiency:

> A Base Case, in which existing policies continue and only limited new actions are considered . . . .
> An Accelerated Supply Strategy, in which the Federal Government takes a number of key actions to increase the domestic supply of energy . . . . ,
> A Conservation Strategy, which would reduce demand for petroleum . . . . ,
> An Emergency Preparedness Strategy . . . .
> Practically speaking, any final Project Independence program would almost surely be a mixed strategy taking elements from each.[7]

The *Project Independence Report* was explicit in its expectation that energy policy would include changes on both the demand side and the domestic supply side of energy markets. An "all of the above" energy policy, as articulated by President Obama, is not new to the United States, but parallels the conclusions of the Ford administration.

The *Project Independence Report* and subsequent policy making in the 1970s created the expectation that there would be initiatives to reduce the net energy imports and that those initiatives would include reductions in the energy intensity of the economy (referred to in the *Project Independence Report* as a "Conservation" Strategy) and would include increases in the domestic production of energy.

The *Project Independence Report* explicitly envisioned energy-efficiency initiatives, stating:

---

7. *Project Independence Report*, US Federal Energy Administration (1974).

	1972	1985	% CHANGE 1972-1985	COMPOUND ANNUAL RATE OF GROWTH 1972-1985
FEA ($11/BBL Import Price)	72.1	102.9	43	2.7%
FEA ($7/BBL Import Price)	72.1	109.1	51	3.2%
Dupree-West (D-W)	72.1	116.6	62	3.8%
Ford Foundation (EPP)				
High Case	72.1	115.0	60	3.7%
Low Case	72.1	93.0	29	2.0%
Nat'l Petroleum Council (NPC)				
High Case	72.1	144.9	101	5.5%
Low Case	72.1	124.9	73	4.3%

Table 3.1. Comparison of Forecasts of Energy Demand, in Quadrillion BTUs

Unlike supply enhancement, where actions are limited by availability of resources, there are innumerable alternative technical possibilities for energy conservation. . . . The FEA Energy Conservation Strategy focused on a limited number of major conservation actions to show the impacts of such a program.[8]

Energy-Consumption Growth Expectations during the Early 1970s

Prior to and soon after the energy crisis, typical expectations were that energy-use growth would continue at rates similar to the historical trend. The *Project Independence Report* summarized prominent contemporary energy-consumption forecasts and compared those to its own.[9] The red oval in Table 3.1 (added here) emphasizes that in the

---

8. Ibid.
9. These appeared in Table III-1 of the Project Independence Report, repeated here as Table 3.1.

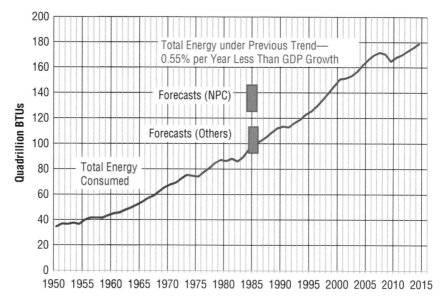

Figure 3.5. Energy Forecasts and Energy Use: Limited-Energy-Efficiency Benchmark

early 1970s, the expectation was that energy use would continue to grow at rates similar to preceding decades.

These forecasts are superimposed on the limited-energy-efficiency benchmark of Figure 3.3 in Figure 3.5. The 1985 energy-consumption forecasts range from slightly below the limited-energy-efficiency benchmark to over 40 quadrillion BTUs above that benchmark. The limited-energy-efficiency benchmark is consistent with or lower than the expectations prevalent in the early 1970s, as represented by the formal forecasts.

Energy Use after the 1973–74 Crisis

The energy crisis led to increased energy prices for oil, natural gas, products substitutable for oil, and products made from oil and other primary sources of energy, such as electricity and gasoline. It motivated an intense period of governmental regulation designed to increase the domestic supply of energy and to reduce the energy intensity of the economy. The crisis brought energy issues to the attention of the general public, including business personnel. This attention remained high until the world oil price crashed in 1986 (see Figure 3.4), after which time the issue of energy imports and energy security began fading.

But the issue of global climate change was coming forward as scientists began warning that combustion of fossil fuels was releasing carbon dioxide into the atmosphere and that the growing concentration of carbon dioxide and other greenhouse gases threatened to substantially warm the earth. In 1979, a National Academy of Sciences report— *Carbon Dioxide and Climate: A Scientific Assessment*—warned of a "global surface warming of between 2°C and 3.5°C." The dangers of global climate change were brought to public attention with the *First Assessment Report* of the Intergovernmental Panel on Climate Change (IPCC) in 1990. Two years later, in 1992, the United Nations Conference on Environment and Development, popularly called the "Earth Summit" was held in Rio de Janeiro.[10] The United Nations Framework Convention on Climate Change, opened for signature at the summit, had as an objective to "stabilize greenhouse gas concentrations in the atmosphere at a level that would prevent dangerous anthropogenic interference with the climate system."[11] And in December 2015, the 21st Conference of the Parties—COP21—also known as the 2015 Paris Climate Conference, reached broad international agreement on climate. Both developed and developing countries have agreed to limit their emissions to keep average global temperature from rising above 2 degrees Celsius.[12]

---

10. http://www.un.org/geninfo/bp/enviro.html
11. http://unfccc.int/essential_background/convention/items/6036.php
12. The bulk of the efforts will involve the energy system, although some efforts will involve avoiding deforestation.

With the growing public and private attention paid to greenhouse gas emissions, environmental consequences of energy use began motivating energy efficiency even as issues of energy imports and energy security faded. Mitigation of carbon dioxide emissions has become a critical public policy issue and can be expected to remain so for the distant future.

Starting with the energy crisis in 1973–74 and continuing through the present, a combination of factors together began a slow cumulative process of energy-efficiency improvements.

The patterns of innovation changed in companies.[13] Companies began to design appliances to use energy sparingly; energy management was integrated into computers; energy-management systems for buildings and enterprises were developed; employees were encouraged to find ways of saving energy; lean manufacturing ideas included energy-use reduction. Reduction in energy use and thus energy cost became an important factor in product and process innovation.

Prior to the crisis, companies were innovating in the products they produced and processes they used to produce and distribute those products, but that innovation was seldom directed toward reducing energy consumption. Innovations *might* reduce energy use, but energy was just one of many inputs. After the crisis, energy costs were seen as important, and there was incentive to direct some innovation toward reducing energy use. As a result, in many organizations, energy-saving technologies and practices that had not been previously envisioned were created and implemented. These changes included the examples illustrated above, as well as many others.

---

13. For a useful overview of energy R&D before 1973, see the recently reissued (in June 2015 from a 1974 book), John E. Tilton, *US Energy R&D Policy: The Role of Economics* (Madison: University of Wisconsin). For a recent overview of the structure of current R&D with a focus on energy, see a 2013 report by the American Energy Innovation Council by J. Rissman and M. Savitz, *Unleashing Private-Sector Energy R&D: Insights from Interviews with 17 R&D Leaders*, http://americanenergyinnovation.org/wp-content/uploads/2013/01/Unleashing -Private-RD-Jan2013.pdf.

There was also significant policy innovation at the federal and state levels. Federal CAFE standards, described above, for cars and trucks were legislated. Federal appliance efficiency standards were developed and implemented. Some states developed mandatory codes and standards. In California, the concept of "decoupling" was initiated in the early 1980s, which allows utility revenues to remain the same, even if the utility sells less of its product (electricity and/or gas) due to efficiency programs. Many of these policy initiatives are discussed in Chapter 6.

As a result of these voluntary individual efforts and governmental policy changes, patterns of capital investment changed. In contrast to the decades before the energy crisis, there was now a profit payoff for capital investments that would reduce energy use, particularly in energy-intense companies. In some areas of the country, utilities became partners with companies, helping to educate them on cost-saving energy efficiency and providing financial incentives to reduce initial investment costs. And even after oil prices dropped in the mid-1980s, there was a risk-management payoff for reducing use of fossil fuels that released carbon dioxide into the atmosphere, given the very real possibility of future carbon dioxide regulations or carbon taxes.

Some companies changed management practices. It became profitable and consistent with the new norms to implement systems to monitor energy use and to hire people responsible for reducing energy use. Larger companies began hiring energy managers with the responsibility for finding ways of reducing energy use. Other behavioral techniques became part of management. Some of these changes were among the examples described above.

The net result: after the energy crisis, the energy-consumption growth rate fell well below the limited-energy-efficiency benchmark, as is shown in Figure 3.6, which graphs the actual energy consumption in the US from 1950 through 2014, with a vertical line drawn at the year 1973.

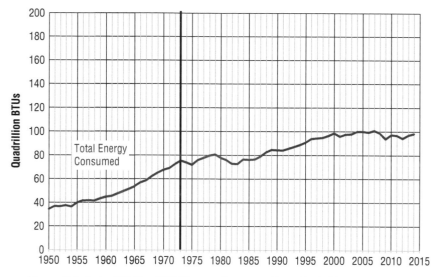

Figure 3.6. Actual Path of US Energy Consumption: 1950–2014

Source: EIA, Monthly Energy Review

The data in Figure 3.1 and in Figure 3.6 are combined into Figure 3.7, which compares the limited-energy-efficiency benchmark with the actual energy-consumption growth. This graph makes it clear just how substantially the energy-consumption growth post-1973 was below the limited-energy-efficiency benchmark. The difference between actual energy consumption and the limited-energy-efficiency consumption benchmark measures the impacts on US energy consumption of enhanced energy-efficiency improvements, as indicated by the green arrow.

Figure 3.7 shows that in 2014 energy consumption would have been about 180 quadrillion BTUs, absent the post-energy-crisis pattern of energy-efficiency improvements, rather than the actual consumption of almost 100 quadrillion BTUs. **Energy efficiency since the 1973 energy crisis has reduced the use of energy by 80 quadril-**

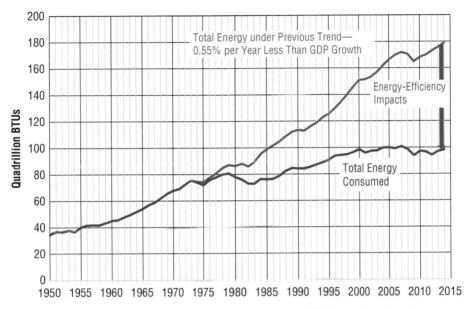

Figure 3.7. Energy Use: Actual vs. Limited-Energy-Efficiency Benchmark

lion BTUs from what it would have been had there not been the cumulative process of energy-efficiency improvements! With the benefit of hindsight, we can see that the role of energy efficiency in shaping the US energy system has been dramatic.

Figure 3.6 and Figure 3.7 also show that US energy consumption in 2014 was the same as in the year 2000, even though the US economy grew 28 percent during those years. The economic downturn in 2008 and 2009 led to reductions in the total energy consumed during those years. But as the economy recovered in 2010 through 2014, the total energy consumption remained relatively constant.

Both the actual energy consumption and the limited-energy-efficiency consumption benchmark data depend on the changing growth rates of GDP in the US economy. But there is another way of examining the changed trends, a way that controls for the changes in

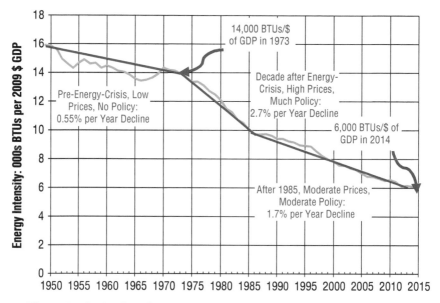

Figure 3.8. Reductions in Energy Intensity of the US Economy

Data source: EIA, Monthly Energy Review

the size of the economy. We can graph the energy intensity of the US economy over time, the ratio of energy consumption to GDP. Figure 3.8 provides that information.

The blue line in Figure 3.8 shows the actual energy intensity of the economy. The three red lines show trends of energy-intensity changes in three different periods.

In the pre-energy-crisis period, there were low energy prices, little or no energy policy initiatives, and low recognition of energy. During this period, as discussed earlier, on average the energy intensity declined 0.55 percent per year.

In the second period, from 1973 through the time of the oil price collapse in 1985, there were high energy prices, many energy policy initiatives, and high recognition of energy use as an important issue. During this period, on average the energy intensity declined 2.7 percent per year.

In the third period, after the oil price collapse, there was a moderate amount of energy policy activity; energy prices were higher than in the pre-energy-crisis period, but lower than in the second period, and there was a reduced, but still very active, recognition of energy as an important issue. During this period, on average the energy intensity declined 1.7 percent per year.

Annual changes of 2.7 percent or 1.7 percent per year may appear to be small. But these changes in energy intensity accumulated over a 40-year period. The net result was that the energy intensity of the US economy reduced from 14,000 BTUs per dollar of GDP in 1973 to 6000 BTUs per dollar of GDP in 2014 (both figures in 2009 dollars), a reduction of 57 percent.

US Domestic Energy Production in the Post-Energy-Crisis Period

As summarized above, the *Project Independence Report* in 1974 identified both energy efficiency and accelerated supply of domestic energy production as key strategic goals, noting that any final Project Independence program would almost surely include both strategies. And President Obama, in his 2013 State of the Union speech, described an "all-of-the-above" approach for further energy progress. The White House Fact Sheet accompanying that speech stated that "renewable electricity generation from wind, solar, and geothermal sources has doubled; and our emissions of the dangerous carbon pollution that threatens our planet have fallen to their lowest level in nearly two decades. In short, the President's approach is working." The President's communication was primarily about increases in domestic *production* of energy, and only secondarily about efficiency in the *use* of energy.

But to this point, I have been attributing the major changes in the US energy situation to the use of energy, to the demand side of energy

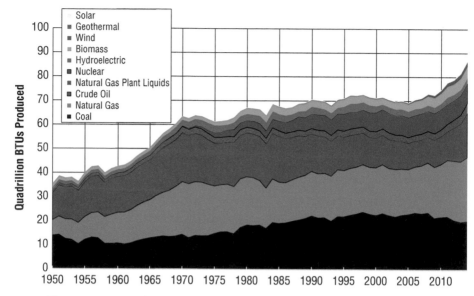

Figure 3.9. Domestic Primary Energy Production: 1950–2014

markets. As shown, the demand side of the energy markets, particularly energy-efficiency changes, fundamentally shaped the US energy system. But what about the domestic production of energy? How much has domestic energy production increased? The answer is in Figure 3.9, which shows domestic primary energy production from 1950 through 2014.

Domestic coal production did increase until relatively recently, when the reduced price and increased availability of natural gas led to large substitutions of natural gas for coal in electricity generation. Natural gas production has increased significantly since 2005, after declining post-1973, with the development of hydraulic fracturing techniques. Crude oil production initially declined, but the decline pattern was reversed since 2005. Nuclear power increased from an insignificant supply in 1973 to a substantial source of energy, although it now is slowly decreasing. Hydroelectricity remained relatively constant during that period, and energy from biomass (primarily ethanol) increased somewhat under legislative incentives.

In total, domestic primary energy production increased from about 62 quadrillion BTUs in 1973 to about 86 quadrillion BTUs in 2014, an increase of about 24 quadrillion BTUs.

Although overall domestic production increased, the domestic production lauded by President Obama—renewable electricity generation from wind, solar, and geothermal sources—remains but a tiny fraction of primary energy.[14] The scale of Figure 3.9 makes it difficult to see the growth of renewable energy supplies. Figure 3.10 expands the vertical scale to show growth of the renewable supplies of energy. (The scale of Figure 3.9 goes from 0 to 100 quadrillion BTUs; in Figure 3.10 the scale goes only one-tenth as far, from 0 to 10 quadrillion BTUs.)

Figure 3.10 shows that about half of the renewable supplies of energy are provided by hydroelectric power and by wood energy. The production of biofuels, wind, geothermal, and solar/PV energy remained nearly zero until 2001. During the last 13 years, these supplies started growing, so that biofuels (dominantly ethanol) and wind energy now both supply about 2 quadrillion BTUs of primary energy equivalent. Geothermal and solar/PV together now supply about half of 1 quadrillion BTUs.

Although these growth rates are high in percentage terms, the total contribution of renewables other than hydroelectric power and wood is small in comparison to the contribution from nuclear power and fossil fuels.

---

14. "The fossil-fuels heat rate is used as the thermal conversion factor for electricity net generation from noncombustible renewable energy (hydro, geothermal, solar thermal, photovoltaic, and wind) to approximate the quantity of fossil fuels replaced by these sources." See http://www.eia.gov/totalenergy/data/monthly/pdf/sec13_6.pdf.

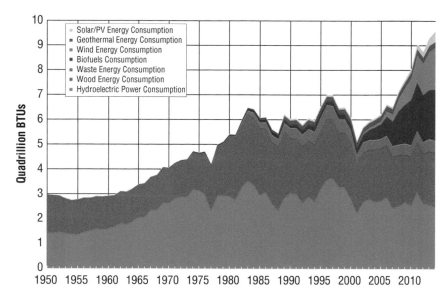

Figure 3.10. Renewable Production of Energy in the US: 1950–2014

Domestic Energy Supply and Energy Demand Together

Since 1973, consumption reductions resulting from energy efficiency have been far greater than the increased domestic supply of all forms of primary energy taken together. This can be seen in Figure 3.11, which plots US energy consumption, domestic supply, and net imports of primary energy from 1959 through 2014. This graph combines production of all the fossil fuels into one, total renewable energy into another, and keeps separate nuclear energy production. In this figure, net energy imports (shown in gray) is the difference between the total energy consumed (shown as a blue line) and total domestic supply of energy (the black, purple, and green areas.)

The dramatic reductions in energy intensity, coupled with the smaller increases in domestic energy production, particularly the very small increases of low-carbon supplies of energy, have fundamentally

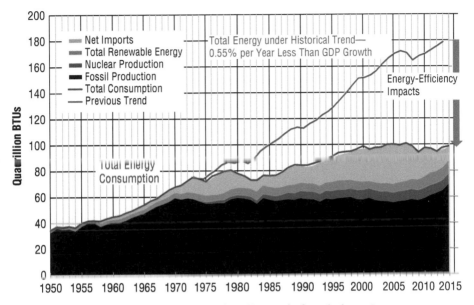

Figure 3.11. US Energy Consumption, Domestic Supply, Imports

improved energy security and have kept carbon dioxide emissions from soaring. The next chapter discusses those impacts.

Although the graph shows that energy from renewable sources remains small, these data are based only on what has already happened. Many of the renewables are now becoming economically competitive, even without the current subsidies, and production from these renewables will be growing much more rapidly than in the past. For renewables, past history is unlikely to be a good indicator of what the future may hold. I firmly expect that will be the case. But any large-scale supply of energy from wind, solar, and geothermal will be a future event; it has not yet happened in the United States.

4

Energy-Efficiency Benefits

ENVIRONMENT AND SECURITY

C hapter 1 pointed out that energy policy typically takes into account growth of the economy, impacts on the domestic and international environment, and on domestic and international security—the energy policy triangle. This chapter quantifies impacts of energy efficiency and energy supply on global climate change and net energy imports.

This chapter shows that reductions in energy intensity of the economy, driven primarily by increases in energy efficiency, have been far more important in reducing the carbon intensity of the US economy, and therefore the carbon dioxide emissions of the United States, than have been all the changes on the supply side of the energy system put together. That is, energy efficiency has been more important than the growth of nuclear power, wind power, solar energy, biofuels, and geothermal energy, all added together.

The last chapter also showed that domestic production of energy has increased by 24 quadrillion BTUs per year and increases in efficiency have reduced the use of energy (from where it would be without enhanced energy efficiency) by 80 quadrillion BTUs per year—well over three times the increase in all forms of domestic energy production. This chapter relates those impacts to net energy imports, showing that energy efficiency has been more important for national security than has the combination of all changes in the supply side of energy markets.

Decarbonization of the US Economy

The role of energy efficiency, and more generally the role of reduced energy consumption, in impacting carbon dioxide emissions can be expressed analytically in terms of a Kaya identity, which expresses carbon dioxide emissions per dollar of GDP as the product of two factors: the amount of carbon dioxide emissions per unit of energy use and the amount of energy use per unit of GDP.[1] Mathematically, this identity can be written as follows:

$$CO_2/GDP = CO_2/\text{Energy Use} \times \text{Energy Use}/GDP$$

The left-hand side of this identity is referred to as the carbon intensity of the economy.

The first term on the right-hand side of this identity represents the supply side of domestic energy markets. Carbon dioxide emissions per unit of energy use is referred to as the carbon intensity of energy consumption. For example, if low-carbon or carbon-free fuels substituted for fossil fuels, and the carbon intensity of energy consumption were reduced by 11 percent, this identity shows that with all else equal, the carbon dioxide emissions would be likewise cut by 11 percent.

The second term on the right-hand side of this identity represents the demand side of US energy markets. It is the energy use per unit of GDP, the energy intensity of the economy. If the energy intensity[2] of the economy were to decline by 50 percent, this identity shows that

---

1. An identity is a mathematical equality that must always be true no matter what the value of the variables. The particular identity, the Kaya identity, is named after Yoichi Kaya. See Yoichi Kaya, in *Environment, Energy, and Economy: Strategies for Sustainability*, eds. Yoichi Kaya and Keiichi Yokoburi (Tokyo [u.a.]: United Nations University Press, 1997). ISBN 9280809113.

2. This is the same as saying "if the energy productivity of the economy were to double." In his 2013 State of the Union address, President Obama set a goal to double energy productivity from the 2010 level by 2030.

with all else equal, the carbon intensity of the economy would likewise be cut by 50 percent. If GDP were not influenced, carbon dioxide emissions would be cut by 50 percent.

The decomposition of carbon intensity of the economy into the two components of energy intensity of the economy and carbon intensity of energy consumption helps in separating the historical impacts of energy efficiency from those of changing energy-supply technologies. And it helps to analyze important differences among policies being proposed or negotiated to reduce greenhouse gas emissions.[3]

Enhancement of energy efficiency reduces the energy intensity of the economy. Provision of low-carbon or carbon-free supplies of energy reduces the carbon intensity of the energy system. Either reduction taken alone reduces the carbon intensity of the economy.

The US Energy Information Administration provides data on carbon intensity of the US economy, energy intensity of the economy, and carbon intensity of energy consumption.[4] The data are graphed in Figure 4.1. The calculation of carbon intensity of energy consumption is illustrated in Appendix B.

The data on carbon intensity of energy consumption is based on energy *consumption*, although for the renewable forms of energy and nuclear energy, domestic production and consumption are equal.[5] The

---

3. For example, in the Paris Conference of Parties (COP21) negotiations, countries did not propose substantial cuts in GDP to reduce carbon dioxide emissions; virtually all proposals were designed to reduce the carbon intensity of their economies.

4. Data source: US Energy Information Administration, Monthly Energy Review, Table 12.1 Carbon Dioxide Emissions From Energy Consumption by Source. That table includes data from 1973 on. I constructed data prior to 1973, assuming that the carbon intensity of three commodities—natural gas, coal, and oil—prior to 1973 had the same values as in 1973. See Appendix B for the calculation method.

5. Carbon dioxide emissions from combustion of imported energy are included, and carbon dioxide emissions from combustion of domestically produced, but exported, energy are not included.

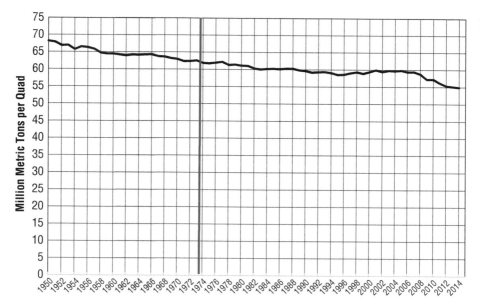

Figure 4.1. Carbon Intensity of US Energy Consumption

carbon intensity of energy consumption[6] has declined from about 63 million metric tons per quad (MMT/Q) in 1973 to 55 (MMT/Q) in 2014. The carbon intensity of energy consumption had been decreasing somewhat more rapidly between 1950 and 1982 because the market share of coal was becoming smaller. Progress, however, stalled from 1982 through 2006. But in the eight years after 2006, carbon intensity of energy consumption declined by 4.4 MMT/Q, primarily because

---

6. The change from 62.6 million metric tons per quad (MMT/Q) to 54.9 was composed of factors that reduced the intensity and other factors that increased the intensity. Reducing the intensity were increased market share of nuclear power (59% of the total change), increased market share of biomass (23% of the total change), reduced market share of petroleum (20% of the total change), increased market share of wind (14% of the total change), increased market share of solar/PV (4% of the total change), and increased market share of geothermal (2% of the total change). *Increasing* the intensity were decreased market share of hydropower (10% of the total reduction), increased market share of coal (10% of the total reduction), and reduced market share of natural gas (2% of the total reduction).

natural gas has substituted for coal and secondarily because the market shares of biomass and wind power have grown.

The data from Figure 4.1, combined with the data from Figure 3.8 (Reductions in Energy Intensity of the US Economy), show the role of the two factors in reducing the carbon intensity of the US economy since 1973. For ease of interpretation, the Kaya identity can be normalized so that each term is expressed as a fraction of its value in 1973. With normalization, the carbon intensity of the economy as a fraction of the carbon intensity of the economy in 1973 is equal to the product of two terms: (1) the carbon intensity of energy consumption as a fraction of the carbon intensity of energy consumption in 1973, and (2) the energy intensity of the economy as a fraction of the energy intensity of the economy in 1973[7] (see Figure 4.2.)

Four data series are plotted in Figure 4.2. The red plot is the carbon intensity of the US economy. Figure 4.2 shows that the carbon intensity has declined by 61 percent: the carbon intensity of the US economy in 2014 was only 39 percent of its value in 1973.

The solid blue line and a solid black line show the two factors that together have led to this decarbonization. The solid blue line is the energy intensity of the US economy. The energy intensity has decreased 55 percent from its 1973 value: in 2014 the energy intensity of the US

---

7. Mathematically, we can write the following two equations where the subscript "yr" can be any year and the subscript "73" indicates the value in the year 1973:

$$(CO_2 / GDP)_{yr} = (CO_2/Energy\ use)_{yr} \times (Energy\ use / GDP)_{yr}$$
$$(CO_2 / GDP)_{73} = (CO_2/Energy\ use)_{73} \times (Energy\ use / GDP)_{73}$$

Dividing the left-hand-side of the first equation by the left-hand side of the second equation and similarly dividing the right-hand side of the first equation by the right-hand side of the second equation gives the following expression:

$$(CO_2 / GDP)_{yr} / (CO_2 / GDP)_{73} = (CO_2 / Energy\ use)_{yr} / (CO_2 / Energy\ use)_{73}$$
$$\times (Energy\ use / GDP)_{yr} / (Energy\ use / GDP)_{73}$$

Each term has a value of 1.0 in 1973. The value of each year is the ratio of the value in that year to the 1973 value.

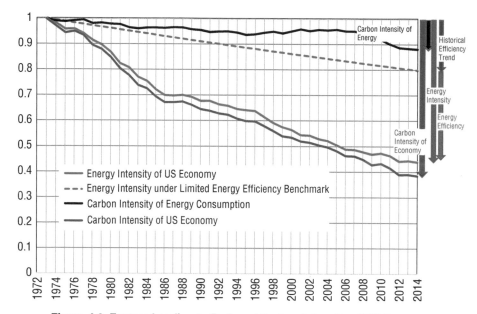

Figure 4.2. Factors Leading to Reduced Carbon Intensity of US Economy

economy was 45 percent of its value in 1973. The black line is the carbon intensity of US energy consumption. The carbon intensity of energy decreased 11 percent: US energy consumption in 2014 was 89 percent as carbon intense as in 1973.

These two factors are multiplicative, as shown in the Kaya identity,[8] leading to the reductions in 2014 carbon intensity of the US economy to slightly less than 40 percent of its value in 1973.

The dashed blue line in Figure 4.2 shows changes in the energy intensity of the US economy if the pre-1973 trends had continued (the energy intensity under the limited-energy-efficiency benchmark). The difference between the dashed blue line and the solid blue line provides an estimate the impact of enhanced energy efficiency on the energy in-

---

8. $40\% = 89\% \times 45\%$.

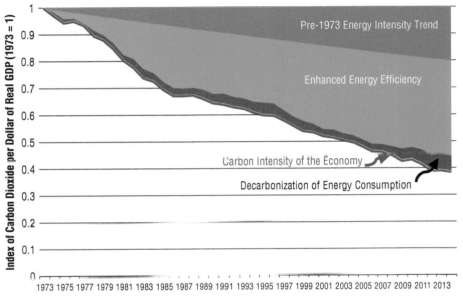

Figure 4.3. Factors Leading to Reduced Carbon Intensity of US Economy

tensity of the US economy. For 2014, the purple arrow and the green arrow show the two components of energy-intensity reductions. The purple arrow represents the historical trends in intensity; the green arrow represents the additional impacts of enhanced energy efficiency since 1973.

The data of Figure 4.2 can be plotted more simply as a cumulative area graph that decomposes the percentage change in carbon intensity of the economy into the three components: pre-1973 trends of energy-intensity reductions (shown in blue), impacts of enhanced energy efficiency since 1973 (shown in green), and impacts of reduced carbon intensity of energy consumption (shown in gray). The data are again normalized relative to 1973 levels, so the carbon intensity of the economy is shown as 1.0 in 1973. These US data are graphed from 1950 through 2014 in Figure 4.3.

The blue plus the green areas in Figure 4.3 show the percentages that energy-intensity reductions have contributed to reduced carbon intensity of the economy, *given* the actual decarbonization of energy consumption. Similarly, the gray area in Figure 4.3 shows the percentage that decarbonization of energy consumption has contributed to reduced carbon intensity of the economy, *given* the actual changes in energy intensity. The red line, the carbon intensity of the US economy over time, here repeats the red line from Figure 4.2.

Since 1973, energy-intensity reductions in the US economy (the green plus the blue areas in Figure 4.3) have been about nine times as important as have reductions in carbon intensity of energy consumption for reducing the carbon intensity of the US economy. Energy-efficiency changes have been about six times as important as reductions in carbon intensity of energy consumption.[9]

These data show that reductions in energy intensity of the economy, driven primarily by increases in energy efficiency, have been far more important in reducing the carbon intensity of the US economy, and therefore the carbon dioxide emissions of the United States, than have reductions in the carbon intensity of energy consumption (e.g., the supply side). That is, over the past 40 years, changes in the economic system that have reduced use of energy have had much greater impact on reducing carbon dioxide emissions (from what they otherwise would have been) than have changes in methods that are employed to produce energy used by the economy. Reductions in the

---

9. The precise partition of energy-intensity reductions is based on the judgment that the historical trends would have continued were there no additional energy-efficiency enhancements. Therefore, the fractions of the energy-intensity reduction associated with enhanced energy efficiency are estimates. But changes in energy intensity are based on objective data collected and published by the US Energy Information Administration.

demand side have been fundamental for addressing sustainability and carbon emissions in particular.[10]

US Net Energy Imports

One particularly important security issue for US energy policy has been the imports, or more precisely the net imports, of energy into the United States. All energy used in the United States must be either produced in the United States or imported from other countries. The net amount of energy imported is the difference between US energy use and US energy production. A reduction in energy use or an increase in US energy production of the same magnitude both have the same impact on reducing net energy imports and thus in improving security.[11]

As noted earlier, the *Project Independence Report* anticipated that changes on both the supply side and the demand side of energy markets would be important for keeping energy imports to acceptable levels.

Previous chapters of this book presented data on US energy production and on energy use. These data support the conclusion that changes on the demand side of energy markets have been far more important components to the reduction of US net energy imports than have changes on the domestic supply side.

---

10. Although the US economy has decarbonized by 61 percent, much more is needed in the United States, as in the rest of the world, to limit global climate change to the 2-degree-Celsius warming recently agreed upon in Paris at the COP21 meeting.

11. This analysis does not consider the specific composition of energy imports or the degree to which different primary forms of energy can substitute for one another in consumption. Energy imports have been dominantly crude oil or refined petroleum products. An increase in oil use from the actual use directly increases oil imports. However, an increase in use of other primary energy from the actual use—say, natural gas—would only indirectly increase oil imports, based on substitution of oil for that primary energy. This analysis does not attempt to quantify those substitutions.

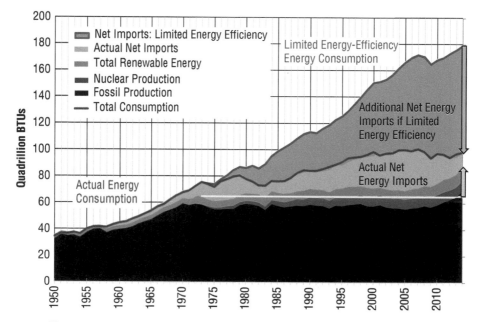

Figure 4.4. US Net Energy Imports: Actual vs. Limited Energy Efficiency

Figure 4.4 shows the actual net energy imports (light gray area) and the net energy imports that would have occurred if the limited-energy-efficiency trends of the pre-energy-crisis period had persisted through 2014 (light gray area plus dark gray area), given the actual levels of domestic production. In that case, there would have been about 90 quadrillion BTUs by 2014, a completely untenable level of imports. The difference between the net energy imports under the limited-energy-efficiency benchmark and the actual net energy imports is about 80 quadrillion BTUs, as shown by the blue arrow.

Figure 4.4 shows that by 2014, fossil fuel production had increased from the 1973 level by about 10 quadrillion BTUs, nuclear energy production by about 8 quadrillion BTUs, and renewable energy production (including wood and hydropower) by about 5 quadrillion BTUs. In total, increases in domestic energy production led to reductions of net energy imports by about 24 quadrillion BTUs, as shown by the gold arrow.

In summary, increases in efficiency have reduced the use of energy by 80 quadrillion BTUs per year—well over three times the increase in all forms of domestic energy production.

Although Figure 4.4 shows the imports that would have occurred absent the enhanced energy efficiency, the United States could not have imported such large quantities of energy. World energy markets could not have sustained such an increase in United States net energy imports. Prices of energy in international trade, particularly oil, would have risen greatly and would have sharply limited net imports.[12]

In addition, the United States would not have been willing to allow such high levels of net energy imports; they would have been politically unacceptable. With such high import levels, the United States would have been extremely vulnerable to interruptions in energy imports. The dependency on the energy imports would have tied US hands in international negotiations, particularly with oil-exporting countries, including those in the Middle East. The great increase of oil prices and oil exports from the Middle East would likely have put even more money in terrorist hands.

These severe economic and security ramifications of massive energy imports would have motivated the United States to make many changes, including greatly increasing domestic production of all primary energy

---

12. If all of the 100 quadrillion BTUs of net imports were oil, the imports would be 46 million barrels per day. The total world production of oil outside of the United States in 2014 was 79 million barrels per day, although the production capacity was greater. World oil markets would not support the United States importing over one-half of the rest of the world's production of oil. Thus this amount of imports of oil would have been impossible or virtually impossible. Other adjustments would be required, including imports of natural gas and coal. Those would have their own economic and environmental problems, and would also face limitations.

sources, but particularly coal and nuclear power, increasing deployment of electric generators, and creating many energy-efficiency initiatives.

The mix of policy responses and economic adjustments that would have been required would have depended on many factors, including the degree to which the various primary energy sources could economically substitute for one another. The mix would depend on important political decisions, involving trade-offs between the three goals of energy policy—improvements in the health and growth of the economy, protection of the domestic and international environment, and enhancement of domestic and international security. Looking backward, what combination of options would have resulted is not at all clear. But what is clear is that the United States would not have been willing, and probably would not have been able, to tolerate the international security implications of such massive energy imports.

Even with the possible policy initiatives and economic adjustments, absent the enhanced energy efficiency there would have been large increases in net energy imports. Quantitative estimates require a more complete analysis of supply responses, of the mix of primary energy forms consumed, and demand substitution options. A complete analysis would be completely speculative and would go well beyond the scope of this book.

However, several observations imply that there likely would have been large increases in net imports of petroleum, coal, and natural gas.

First, much of the energy efficiency has been in the transportation sector, a sector almost completely reliant on petroleum. And because electric vehicles have only recently become available, there would have been relatively little ability to substitute away from oil for cars, trucks, buses, or airplanes. For those reasons alone, it is most likely that petroleum imports would have increased greatly.

The United States can import only a limited amount of electricity from Mexico and Canada. Therefore, the large increases in electricity use would have required large increases in electricity generated in the United States. The great increases in electricity generation would have required construction of large numbers of new electric generation stations—most likely coal-fired and nuclear, but possibly natural gas—and increases in transmission lines. More water would have been used for cooling towers in those generation stations. More particulates would be released into the air from the coal-fired units. The United States would have had to burn much more coal and probably more natural gas. The United States would have been importing large amounts of natural gas in liquefied form (LNG), rather than exporting LNG, as it does now. Natural gas would be very expensive, and chemical companies would have been leaving the United States rather than moving back. More coal would be mined in the United States, and probably more would be imported. More unit trains carrying coal would be in operation.

The changes in imports of oil, natural gas, and coal would have led to important national security risks and increased damages to the local and international environment. But fortunately, those detrimental impacts never came to pass: the combination of some increased energy production and much energy efficiency has allowed the United States to avoid facing those problems. Now we can look back and say, "Whew! We dodged a bullet!" Or maybe a cruise missile.

5

Sectoral Disaggregation
of Energy Consumption

I n this chapter, we turn from economy-wide trends to trends in major
energy-consuming sectors of the economy. This examination shows
that the enhanced energy efficiency discussed previously in this book
has not been concentrated in any single part of the economy, but rather
has been broadly distributed throughout all of the major energy-
consuming sectors. Energy efficiency is literally everywhere.

Industrial, Transportation, Residential, Commercial Sectors

Figure 5.1, using data from the US Energy Information Administration,
separates energy consumption into the four major sectors: industrial,
transportation, residential, and commercial. The vertical line at 1973
marks the beginning of the 1973 energy crisis.

For each of the four sectors, the trends in energy-consumption
growth changed significantly soon after the beginning of the energy crisis; there were very pronounced inflection points. It also shows that
the largest post-1973 change from the trend—the most pronounced
inflection—was in the industrial sector.

Figure 5.2 superimposes onto Figure 5.1 trend lines for consumption in each of the four sectors. The four broken lines extrapolate the
pre-energy-crisis trends in consumption growth for each of the sectors.
A comparison of these limited-energy-efficiency trend lines with the
actual changes in energy consumption for the sectors makes it clear
that actual energy consumption began diverging from the historical

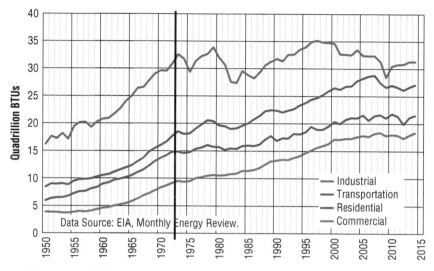

Figure 5.1. US Energy Use by Major Consuming Sector

Source: EIA, Monthly Energy Review

trends soon after 1973 but the time pattern and the degree of divergence varied among the sectors:

- in both the industrial and the residential sector, there was a sharp divergence that began immediately after 1973;
- in the commercial sector, there was a significant divergence that began immediately after 1973;
- in the transportation sector, there was a large divergence that did not begin until 1980.

For all four sectors, increased prices, the expectation of further price changes, awareness of energy issues, and development of new policies to encourage (or mandate) efficiency started to have immediate impacts and these impacts grew over time, leading to the growing

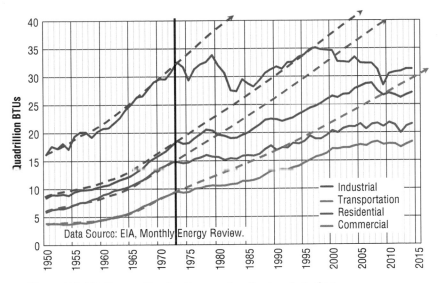

Figure 5.2. US Energy Use by Consuming Sector: Actual vs. Limited-Energy-Efficiency

Source: EIA, Monthly Energy Review

divergence. New and more energy-efficient technologies and practices were adopted over time, leading to even more of a divergence.

In the transportation sector, there was only a relatively small immediate impact on automotive vehicle miles or passenger miles traveled. But as discussed in Chapter 2, the CAFE standards on automobile fuel efficiency did not come into effect until 1978 for cars and 1979 for trucks, and then only for the newly purchased vehicles. Once CAFE standards came into effect, the divergence between past trends and the actual energy use patterns grew quickly.

The industrial sector is more complicated. The divergence was much sharper than for the other sectors. For this sector, in addition to the changes in technologies and practices, there were important shifts in the structure of US industry. Some of the industrial sector divergence shown in Figure 5.2 was the result of shifts in the structure of industry, associated with changing international trade patterns. These changes continued after the energy crisis, but the magnitude of the shifts accelerated after the year 2002, when China started becoming the world center for heavy manufacturing. Industrial restructuring has been a significant component of the reductions in industrial sector energy intensity. We will return shortly to that issue.

However for the other three sectors—transportation, residential, and commercial—international trade issues had little or no effect on energy-consumption patterns. These are activities that generally were not outsourced to other countries.[1] Given the international trade impacts on the industrial sector and the lack of such impacts on the other three sectors, it is useful to reconstruct Figure 3.8 (Reductions in Energy Intensity of the US Economy) separating energy per unit of GDP for the industrial sector from energy per unit of GDP for the aggregate of the other three sectors.[2] Those data are plotted in Figure 5.3.

In Figure 5.3, for both the industrial sector and the aggregate of the other three sectors, the rate of energy intensity reduction varies among the three time periods. In the pre-energy-crisis years, energy use per unit of GDP was declining relatively slowly in the industrial sector—0.8 percent per year—and was declining very slowly, if at all, in the aggregate of the other three sectors (residential, commercial, transportation). For the aggregate of the other three sectors the average rate of decline during these years was 0.3 percent year, but the ratio

---

1. Some commercial activities are now being outsourced to other countries, but this has not been happening in large amounts until quite recently.

2. These are *not* ratios of the energy use in a sector to some measure of activity in the sector, but ratios of energy used in these sectors to the *total* GDP of the economy. Therefore the two intensity measures, when added together, equal the total intensity of the entire economy, shown in Figure 3.8.

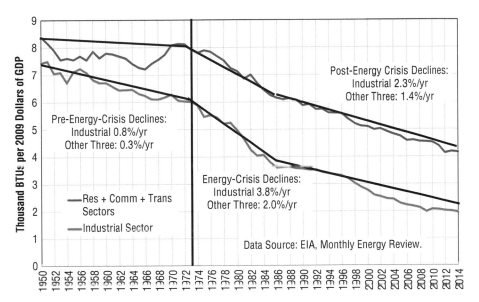

Figure 5.3. Reductions in Energy Use per Dollar of GDP: Industrial and Other Three Sectors

Source: EIA, Monthly Energy Review

increased in some years and decreased in others. One could interpret this time interval as one in which there was no systematic reduction in energy use per dollar of GDP for the aggregate of the other three sectors.

In the second period, from 1973 through the 1985 oil price collapse—a time of high energy prices, many energy policy initiatives, and high recognition of energy use as an important issue—the average rate of decline in the industrial sector was 3.8 percent per year and in the aggregate of the other three sectors was 2.0 percent.

In the period after the oil price collapse (roughly 1986 to the present)— a time of moderate federal energy policy activity but continued state policy initiatives, energy prices higher than in the pre-energy-crisis period, but lower than in the second period immediately after the crisis, and reduced (but still important) recognition of energy as an important issue—the average rate of decline was reduced to 2.3 percent per year in the industrial sector and 1.4 percent per year in the aggregate of the other sectors. In the industrial sector, the decline rate for these years was not uniform, as can be seen in Figure 5.3.

We can compare actual energy consumption with energy consumption under the limited-energy-efficiency benchmark for the industrial sector and for the aggregate of the other three sectors (residential, commercial, and transportation). These two comparisons are shown in Figure 5.4, which is analogous to Figure 3.7 (Energy Use: Actual vs. Limited-Energy-Efficiency Benchmark). This graph labels the differential from the trend line for the industrial sector as "Industrial Energy Efficiency Impacts with Structural Shifts," recognizing that a significant amount of the industrial sector reductions in energy consumption from the pre-1973 trend represents important structural shifts in the US industrial sector associated with changing international trade patterns.

Figure 5.4 shows that for the aggregate of residential, commercial, and transportation sectors, the energy-efficiency impact grew to roughly 35 quadrillion BTUs by the year 2014. For the industrial sector, energy-efficiency impacts with structural shifts grew to roughly 45 quadrillion BTUs by the year 2014.

Structural Shifts and the Industrial Sector

Structural shifts are particularly significant for the industrial sector because of the movement of manufacturing to China.[3] But even though

---

3. This point has particular relevance for examining global climate change issues. If the particularly carbon-intensive products are manufactured in China and exported to the United States, then although the carbon emissions in the United States decrease, the US decreases are

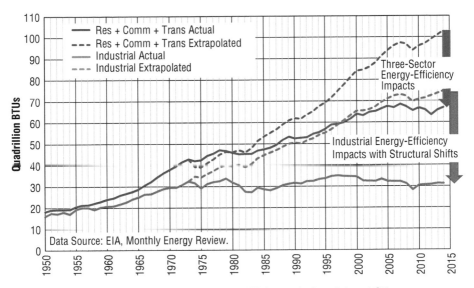

Figure 5.4. Actual vs. Limited-Energy-Efficiency: Industrial and Other Three Sectors

Source: EIA, Monthly Energy Review

heavy manufacturing moved to China, the industrial sector in the United States did not collapse, although the growth of industrial production slowed very importantly after the year 2000. Figure 5.5 shows the Federal Reserve Industrial Production Index[4] and the energy use from 1950 to 2014, and illustrates the continued growth of industrial production. The index of industrial production continued climbing at

---

matched by roughly equivalent Chinese increases in emissions, or emissions may rise if China uses more carbon-intense processes. However, this point has little or no relevance for energy security. Even if the goods are produced in China, the US net energy imports do decrease, improving energy security. Possible short-term disruptions in the supply of consumer goods imported from China do not have the same security implications as do short-term disruptions in the supply of energy.

4. The industrial production index measures real output of all manufacturing, mining, and electric and gas utility establishments in the United States. See http://www.federalreserve.gov /releases/g17/IpNotes.htm for details about this index.

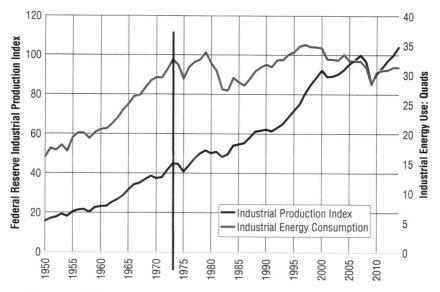

Figure 5.5. US Industrial Production and Energy Use

Source: EIA, Monthly Energy Review; Board of Governors, Federal Reserve System (US)

roughly the same rate up until the year 2000, after which time the growth rate declined and then the index itself decreased.[5] However, industrial sector energy consumption fluctuated between 1973 and the year 2000 with year 2000 consumption being roughly equal to energy consumption in 1973, 27 years earlier.

Although industrial production continued to grow, after 1973 there was a shift away from energy intensive manufacturing and other industrial production and toward much less energy-intense products and processes. This structural shift has been important in reducing energy intensity of the industrial sector.

_____

5. The sharp decline in the growth of industrial production is related to the growth in China as a manufacturing center for the world.

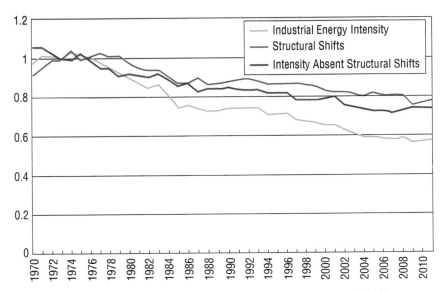

Figure 5.6. Industrial Sector Energy Intensity: Role of Structural Shifts

Source: US Department of Energy, Office of Energy Efficiency and Renewable Energy (EERE)

The US Department of Energy, Office of Energy Efficiency and Renewable Energy (EERE), estimates that slightly less than half of the change in industrial sector energy intensity since 1973 was the result of structural shifts. The EERE analysis does not attempt to estimate what fraction of the structural shifts was the result of international competition and what fraction was the result of deliberate attempts to shift from energy intensive products. The EERE data are plotted in Figure 5.6. Data are normalized so that each intensity measure is shown as a fraction of its 1973 value.[6]

Whether one includes that structural shift within the quantification of energy efficiency depends on the definition one uses. Throughout

6. Data for this graph are from http://www1.eere.energy.gov/analysis/eii_trend_data.html.

this book I have used a broader definition of energy efficiency than the one used by EERE in this study.[7] This book uses this concept: energy efficient changes are economically efficient reductions in energy use. Energy-efficient changes reduce the use of energy but do not reduce the overall value of goods and services available to the US economy.

Under the definition used within this book, almost all of the structural shifts since 1973 are included as energy efficiency. The impact of energy efficiency on the economy by 2014 is almost 80 quadrillion BTUs. Under the concept used by EERE, at least in that particular study, only one half of the 45 quadrillion BTU industrial sector energy use reductions (about 23) would be counted as energy efficiency. Under the EERE concept, the total impact of energy efficiency on the US economy since 1973 could be estimated as 57 rather than 80 quadrillion BTUs.

In the rest of this book, I include structural shifts as energy-efficiency adjustments, so this book continues to use the 80 quadrillion BTUs estimate. Some readers will prefer the EERE concept and can adjust the various estimates of energy-efficiency impacts accordingly.

Energy Efficiency and the Rebound Effect

This book has shown that energy efficiency in the United States has been a dominant force shaping energy imports and greenhouse gas emissions. Recently, however, a much publicized study[8] and high-profile essays[9] from The Breakthrough Institute have challenged the idea that

7. The concept EERE uses is "Energy efficiency refers to the activity or product that can be produced with a given amount of energy; for example, the number of tons of steel that can be melted with a megawatt hour of electricity." See http://www1.eere.energy.gov/analysis/eii _efficiency_intensity.html. This concept is closer to the definition from physics, as discussed in Chapter 1 of this book.

8. See J. Jenkins, T. Nordhaus, and M. Shellenberger, *Energy Emergence: Rebound and Backfire as Emergent Phenomena*, report by the Breakthrough Institute (2011), http://thebreakthrough .org/blog/Energy_Emergence.pdf.

9. For example, see M. Shellenberger and T. Nordhaus, "The Problem with Energy Efficiency," OpEd, *New York Times*, October 8, 2014, A35.

energy efficiency can be expected to lead to significant reductions in energy use, particularly when viewed over an extended period of time. The study states:

> Economists, however, have long observed that increasing the efficient production and consumption of energy drives a rebound in demand for energy and energy services, potentially resulting in greater, not less, consumption of energy. Energy productivity improvements over time reduce the implicit price and grow the supply of energy services, driving economic growth and resulting in firms and consumers finding new uses for energy (e.g., substitution). This is known in the energy economics literature as energy demand 'rebound' or, when rebound is greater than the initial energy savings, as "backfire."[10]

The rebound effect is an important issue for those working on the details of energy-efficiency policy design and should not be ignored.[11] Yet popular discussions of the rebound effect, such as that cited above, often overstate the issue. In particular, the claim that at an economy-wide level, rebound effects are either very high or can lead to "backfire," has no valid empirical support in developed economies and remains speculative in developing economies.[12]

An example of the rebound effect was incorporated into Figure 2.6 (Energy Used by Cars and Light Trucks) in the context of automobile fuel efficiency. For cars and light trucks, an increase in miles-per-gallon fuel use (say by 10 percent) would reduce the cost per mile of driving, and people would drive more. If people drive 2 percent more, then the

---

10. Ibid., p 4.

11. Rebound effects have been incorporated into US governmental analyses of energy-efficiency impacts as early as for the 1976 *National Energy Outlook*, published by the Federal Energy Administration. See pages C-12, C-13, showing vehicle miles traveled of automobiles as a decreasing function of the cost per mile of driving. The cost per mile of driving decreases when the miles per gallon of the automobile fleet increases.

12. See Danny Cullenward and Jonathan G. Koomey, "A Critique of Saunders' 'Historical Evidence for Energy Efficiency Rebound in 30 US Sectors,'" *Technological Forecasting and Social Change* 103 (2016): 203–13.

total reduction in fuel use in automobiles would be approximately 8 percent, a 10 percent reduction because of the increased miles-per-gallon and a 2 percent increase because people drive more. These rebound effects can be expected to partially offset some of the energy use reductions associated directly with the more efficient technology.

However, the total energy impact is typically far more complicated. If people drive 2 percent more, they may choose to travel by airplane less, reducing use of fuel for air travel. This shift from air transportation implies that the 8 percent reduction in fuel use in automobiles would understate the total reduction in energy use. The shift might even more than compensate for the 2 percent increase in fuel use because people drive more. In that case, the adjustments taken together would *not* partially offset some of the energy-use reductions associated directly with the more efficient technology.

But there are more complications. If consumers save 8 percent on the cost of buying gasoline for cars, they are likely to spend a large fraction of this saving on other goods. If those other goods are produced in the United States, US production would increase, requiring more energy, and benefitting the economy if the US productive resources were below full employment. If the US economy were at full employment, the additional production of those goods would require resources to shift from other production, leading to a reduction in the production of and energy use for still different goods. If the other goods are produced in another country, production would increase there, using more energy; energy would be used to transport those goods to the United States. Many attempts to quantify the final results of all adjustments together—including the primary study behind the Breakthrough Institute's most dramatic findings—have been too flawed to give any meaningful information.

It is important to note that whatever the impact on net energy consumption, the rebound effect is beneficial for those whose behavior adjusts. The 2 percent increase in driving in the example above would occur because the drivers saw a personal benefit, above the additional cost of more driving. The rebound would be the result of them making choices that made them better off. The expenditure on other goods would occur because the purchases were worth more to consumers than their cost. The additional production of these other goods occurs because it is profitable for firms to increase their production. For individuals adjusting, the rebound effect is welfare enhancing.

But for society as a whole, the rebound effect can be either beneficial or harmful. Absent unpriced externalities, the rebound effect is beneficial for the aggregate of society. But for energy, the unpriced externalities of environmental impact and national security effect could be large enough to exceed the individual benefit, at least in developed nations. And for the automobile example, the societal cost of additional highway congestion also could exceed the individual benefits. So these welfare effects are also complicated.

The question for policy is how large can we expect rebound effects to be and how large have they been historically.

For some household and commercial uses of energy, for example, refrigeration, energy use is on a fixed cycle, so it is unlikely that consumers change the cycle as refrigerators became more efficient. One could speculate that consumers have purchased more refrigerators as a result of the reduction in energy cost. The most recent Residential Energy Consumption survey shows that almost every home has a refrigerator, and less than 25 percent of homes have two or more. But each new refrigerator uses less than one-quarter of the electricity of a 1973 refrigerator. Thus, if there is any rebound effect through increased ownership of refrigerators, the increased ownership has not offset much of the increase in efficiency.

For other household or commercial uses, people directly control the usage. They decide at what temperature to set their thermostats for heating and cooling; they turn lights, computers, televisions, and stoves on and off. For these uses, the rebound effect can be larger, although generally limited. Homes and businesses generally are not heated above comfort levels, no matter how energy efficient the heating system. Businesses installing motion-sensing/light-sensing switches or installing LEDs are unlikely to compensate by lighting offices more brightly. No empirical evidence exists that these effects lead to other than a small rebound, say in the range of 10–30 percent of the change in efficiency of the appliances.

For automobile transportation, the rebound effect is likewise small. Fuel efficiency of light-duty vehicles roughly doubled since 1973, and the real gasoline price in 2002 was approximately the same as the 1970 price before the oil crisis.[13] But Figure 2.5 (Vehicle Miles Traveled [trillions]: All US Roads) shows that the growth of vehicle miles traveled slowed down slightly; it did not accelerate. There is no evidence of a substantial rebound, much less of "backfire."

For air transportation, the magnitude is less certain, but the rebound could not have been large. Figure 2.12 (Energy Intensity of Certified Air Carriers) showed that the energy use per passenger mile was reduced by a factor of roughly four. Figure 2.13 (Airline Passenger Miles and Fuel Use) shows that airline travel has in fact grown, but much of that growth could be the result of growing income. But Figure 2.13 also shows that the total use of fuel by certificated carriers has been declining, not growing rapidly as would have been the case with "backfire."

In industry, the uses of energy are so heterogeneous that it is difficult to characterize rebounds. However, the data of this chapter shows

---

13. http://www.eia.gov/totalenergy/data/annual/showtext.cfm?t=ptb0524. In 1973, the average price for leaded regular gasoline was $1.467 (all data in 2005 dollars); in 2002 the price was $1.473 for unleaded regular. In later years, gasoline price increased sharply, to $3.11 in 2011.

that energy use in the industrial sector has not grown since 1973, even though industrial sector output has continued to grow.

Examining the aggregate use of energy in the United States shows that there could not have been large rebound effects in aggregate. In Chapter 3 of this book, it was shown that the US economy-wide energy intensity decreased by 57 percent from 1973 to 2014 and that energy efficiency has limited energy use to 100 quadrillion BTUs per year rather than the 180 quadrillion BTUs per year projected absent enhanced energy efficiency. In Chapter 5, it was shown that energy efficiency reduced energy use from its previous trends in every consuming sector of the economy. These data are inconsistent with the claim that energy efficiency is likely to increase energy use or even that rebound effects have sharply limited energy reductions. However, these aggregate data, while ruling out "backfire" as a common phenomenon, do not allow a good quantification of aggregate rebound effects. Direct inference of the magnitude of rebound effects is precluded because total energy-efficiency gains have resulted from a combination of many forces acting simultaneously—energy price, technology advances, regulatory policies, and individual expectations. But these data on their face are inconsistent with the claim that rebounds have been large and even more inconsistent with the claim that there has been "backfire" in the overall US economy.[14]

Although in developed economies the possibility of large rebounds or even "backfire" is remote, some aspects of the net contribution

14. More literature on the rebound effect is available: Cullenward and Koomey, "A Critique of Saunders' 'Historical Evidence for Energy Efficiency Rebound in 30 US Sectors,'" 203–13; Kenneth Gillingham, David Rapson, and Gernot Wagner, "The Rebound Effect and Energy Efficiency Policy," *Review of Environmental Economics and Policy* 10, no. 1 (2016): 68–88; Severin Borenstein, "A Microeconomic Framework for Evaluating Energy Efficiency Rebound And Some Implications," *The Energy Journal* 36, no. 1 (2015): 1–21; Inês Azevedo, "Consumer End-Use Energy Efficiency and Rebound Effects," *Annual Review of Environment and Resources* 39 (2014): 393–418.

of energy efficiency do need more research, particularly in developing economies. Energy efficiency can lead to some combination of energy reduction and increased economic activity—increased economic growth—when there is substantial unemployment of labor and other resources. In developed economies, such as the United States, the greatest impact is likely to be reduced energy use. However, in developing countries with substantial resource unemployment, if there is rebound—and there is little empirical evidence on this one way or another in such countries—such rebounds would contribute to economic welfare and development. That is, if industrial and commercial activity increases as a result of energy efficiency, through a rebound effect, then energy efficiency may be a way to deliver increased welfare without decreasing energy use or associated carbon dioxide emissions in such countries. More empirical work on the rebound effect is clearly needed in less developed countries.

We turn now to additional policy issues of amplifying energy efficiency.

6

Amplifying Energy Efficiency

What enabled such large enhancements in energy efficiency? A large portion of the energy-efficiency innovation, new technology creation, enhanced technology implementation, and changed company practices would have occurred through normal market processes motivated by the post-1973 increased energy prices and the greater awareness of energy issues.

But in addition to these market processes were institutions, practices, communications, regulations, and policies that amplified energy efficiency: these sped up the rate of energy-efficiency changes and increased the depth of those changes. Some of these were developed as direct responses to the many barriers discussed in Chapter 1 (Table 1.1. Some Barriers to Energy-Use Optimality) that inhibit full implementation of energy efficiency.

These amplifying factors can be seen as part of broad societal and economic forces which changed the United States from an economy and a culture that almost completely ignored energy efficiency to one in which energy-efficiency gains have become a continuing process.

Below are some examples that combine some of these interwoven factors to amplify energy efficiency.

Information/Labeling/Nudges

For decades, the federal government has been making an important investment in broad collection and provision of energy information. Although prior to the energy crisis of 1973 energy information was limited and scattered, now energy information is readily and broadly accessible. The Energy Information Administration (EIA), part of the US Department of Energy, "collects, analyzes, and disseminates independent and impartial energy information to promote sound policy making, efficient markets, and public understanding of energy and its interaction with the economy and the environment."[1] This book could not have been written without the high-quality data made available by the EIA; much energy-consumption data in this book have been gathered and made available on the EIA web pages.

EIA is a direct successor to the offices within the US Federal Energy Administration, itself created under the Federal Energy Administration Act of 1974 as a direct response to the 1973–74 energy crisis.

In addition to the EIA, after the energy crisis, the federal government, in particular the Department of Energy (DOE), the Federal Trade Commission (FTC), the Environmental Protection Agency (EPA), and many utilities and states launched information programs to encourage energy-use reduction.

The best-known and most successful effort in this area is the ENERGY STAR program, run by the US EPA and DOE. The EPA established this voluntary program in 1992, and Congress further authorized it under the 2005 Energy Policy Act "to identify and promote energy-efficient products and buildings in order to reduce energy consumption, improve energy security, and reduce pollution through voluntary labeling of or other forms of communication about products and buildings that meet the highest energy-efficiency standards."[2]

The maker of a product can have it certified as ENERGY STAR and can display the ENERGY STAR label (Figure 6.1) if it meets the energy-efficiency requirements set forth in ENERGY STAR product

1. http://www.eia.gov/about/
2. https://www.energystar.gov/about/

Figure 6.1. Energy Star Logo

Source: https://www.energystar.gov/about/

specifications. EPA establishes these specifications based on the following set of key guiding principles:

> Product categories must contribute significant energy savings nationwide and must deliver features and performance demanded by consumers. If the certified product costs more than a conventional, less-efficient counterpart, purchasers will recover their investment in increased energy efficiency through utility bill savings, within a reasonable period of time.[3]

The ENERGY STAR program and logo provide information to influence consumers toward purchasing those products certified to be ENERGY STAR. But to gain that competitive advantage, a product

---

3. https://www.energystar.gov/products/how-product-earns-energy-star-label

must meet ENERGY STAR standards. Thus there is an incentive for the producers to improve energy efficiency of their products in order to gain competitive advantage. The requirement also provides incentives for producers to provide a range of products embodying different degrees of energy efficiency, including some products just meeting the regulatory requirement and others just meeting the requirements for ENERGY STAR certification.

Research by Sébastien Houde shows "[s]ome consumers appear to rely heavily on Energy Star and pay little attention to electricity costs, others are the reverse, and still others appear to be insensitive to both electricity costs and Energy Star. . . . Energy Star influences firms' decisions. Focusing on the refrigerator market, firms offer products that bunch exclusively at the minimum and Energy Star standards."[4]

The FTC requires EnergyGuide labels to be displayed at the point of retail sales for most appliances: boilers, central air conditioners, clothes washers, dishwashers, freezers, furnaces, heat pumps, pool heaters, refrigerators, televisions, water heaters, and window air conditioners. Figure 6.2 shows a typical EnergyGuide label.[5] Prominent on the label is information on the estimated yearly operating cost, the yearly operating cost range of similar models, the estimated yearly use of electricity, and the ENERGY STAR logo if the product has qualified as an ENERGY STAR device.

The EnergyGuide label encourages consumers to pay attention to the operating costs of their potential new appliance purchases. Knowing this, manufacturers have an increased incentive to engineer their products to reduce energy costs, where such reductions are cost-effective. The ENERGY STAR voluntary labeling program has also

---

4. Sébastien Houde, "Managing Energy Demand with Information and Standards," 2012,. http://purl.stanford.edu/ys203dx0462, page p. 4.
5. http://www.consumer.ftc.gov/articles/0072-shopping-home-appliances-use-energyguide -label

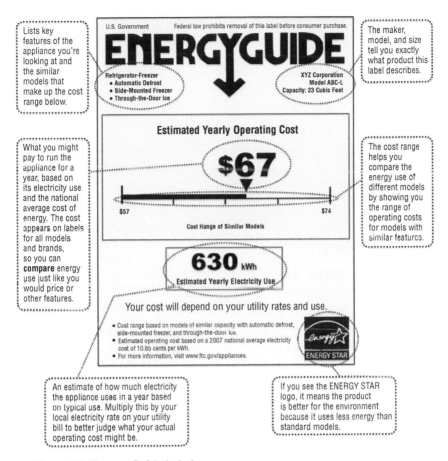

Figure 6.2. Energy Guide Label

Source: www.consumer.ftc.gov/articles/0072-shopping-home-appliances-use-energyguide-label

worked to change overall industry efficiency that in turn has allowed
DOE to adopt minimum energy performance standards (MEPs).[6]

---

6. More detail on the origins of the EnergyGuide label is contained in M. Taylor, C. A.
Spurlock, and H-C. Yang, (2015) *Confronting Regulatory Cost and Quality Expectations: An
Exploration of Technical Change in Minimum Efficiency Performance Standards,* Lawrence Berkeley
National Laboratory 1000576. The 1975 Energy Policy and Conservation Act (EPCA, Pub. L.

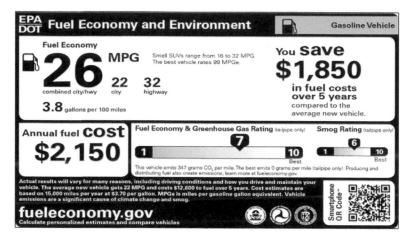

Figure 6.3. New Car Fuel Efficiency Label

Source: http://www.fueleconomy.gov/feg/Find.do?action=bt

Similarly, new cars and light-duty trucks must carry labels with the fuel economy shown in terms of miles per gallon and in terms of gallons per 100 miles. Not only is there an estimate of the annual fuel cost but also a comparison of fuel costs with the average of new vehicles. The newest version of automobile fuel economy labels[7] is shown in Figure 6.3.

Labels for appliances or for automobiles make it easy for consumers to compare energy use among competing products. Importantly, the prominent display of estimated annual operating cost brings con-

---

No. 94-163) introduced three of the four main features of US federal energy-efficiency policies for appliances and other equipment: test procedures, energy-consumption labeling, and minimum efficiency performance standards (MEPs). The EPCA directed the head of what is now the DOE to direct the National Bureau of Standards (now the National Institute of Standards and Technology) to develop test procedures. The EPCA also directed the FTC to develop and administer a mandatory energy labeling program covering major appliances, equipment, and lighting, with the first appliance labeling rule established in 1979 and all covered products required to carry the label starting in 1980.

7. http://www.fueleconomy.gov/feg/Find.do?action=bt1

sumer attention to energy costs rather than only to the purchase price. Labels make it easy for consumers to make economically attractive choices. This class of programs can be seen as a direct response to the market failure and the behavioral issues—"Poor information about electricity use"—identified in Table 1.1 (Some Barriers to Energy-Use Optimality).

Fuel economy labels for automobiles work in conjunction with the CAFE standards. By encouraging consumers to take into account fuel costs of the new vehicles, they provide a motivation to purchase more fuel-efficient vehicles and thus, taken alone, would increase the market share of fuel-efficient cars and trucks. This shift in market shares makes it easier for automobile manufacturers to meet the CAFE standards.

For both new buildings and many existing buildings, certification and labeling is available through the LEED certification program run by the non-governmental organization the US Green Building Council[8] (see Figure 6.4). The rating system has been developed to promote design and construction practices supportive of environmental practices and supportive of occupant health. Launched in 2000, LEED initially had one rating system for new construction. It now has several rating systems for new building construction, existing building operation, interior design, and neighborhood development. The certification level depends on credits earned in site selection, energy efficiency and carbon dioxide releases, water efficiency, sustainable materials, and occupant health. Energy efficiency is only one of many criteria and thus may not be a particularly important characteristic of a LEED-certified building.

There have been questions as to whether LEED certification has actually led to construction of more energy-efficient buildings. For

---

8. The 'LEED Certification Mark' is a registered trademark owned by the US Green Building Council, and is used by permission.

Figure 6.4. LEED Certification Mark

Source: http://www.greenedu.com/storage/leed-ga
-downloads/10-USGBC%20Logo%20Guidelines.pdf

example, a 2012 *USA Today* story reports that its survey of "LEED-certified commercial buildings shows that designers target the easiest and cheapest green points."[9] According to the US Green Building Council, 92.2 percent of LEED-certified new construction projects "are improving energy performance by at least 10.5%, according to an analysis of 7,100 projects."[10] Although research has not shown definitively the degree to which LEED certification has led to more energy-efficient buildings, this rating, certification, and labeling system does provide a strong motivation to construct buildings consistent with the points allowable under the certification rules in effect. This weighting of the various points continues to evolve over time, with increasing emphasis on energy and water efficiency.

In contrast to the LEED program, the ENERGY STAR new homes program, created in 1995, focuses sharply on energy efficiency rather than on a wide range of non-energy characteristics. According to the EPA, "a new home that has earned the ENERGY STAR label has

9. http://www.usatoday.com/story/news/nation/2012/10/24/green-building-leed
-certification/1650517/

10. http://leed.usgbc.org/bd-c.html

Figure 6.5. Energy Star Certified New Home Label
Source: http://www.energystar.gov/index.cfm?c=new_homes.hm_index

undergone a process of inspections, testing, and verification to meet strict requirements set by the U.S. Environmental Protection Agency (EPA), delivering better quality, better comfort, and better durability"[11] (see Figure 6.5). This whole home program provides unambiguous incentives for builders to offer more energy-efficient homes on the market and provides information for consumers considering purchasing a new home.

Another type of behavioral nudge is used by firms like Opower and its competitors, working with utilities and other energy providers. For example, a utility may include a "Home Energy Report" (developed by Opower) insert in a customer's monthly bill, providing a number of behavioral nudges. The insert compares the customer's energy consumption with that of all neighbors and of energy-efficient neighbors. Low energy consumption earns a smiley face; large energy consumption relative to neighbors merits a frowny face. The Opower insert provides no financial incentives to the customers whatsoever. But it does

---

11. http://www.energystar.gov/index.cfm?c=new_homes.hm_index

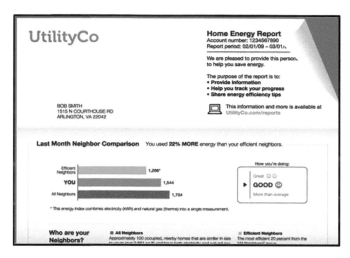

Figure 6.6. Opower Home Energy Report

Source: https://opower.com/solutions/energy-efficiency

provide tips on reducing energy use. An Opower electricity bill insert[12] is shown in Figure 6.6.

This simple behavioral nudge from Opower reduces electricity consumption on average by about 2 percent, with the greatest percentage electricity reduction for those homes that use more electricity than average. This innovation in providing energy information has become widespread because a key policy change has also occurred: public utilities commissions have begun to recognize "behavioral savings"—changes in behavior of energy users—as a legitimate category of energy efficiency that can be funded through utility programs. Thus many state regulatory commission policy decisions have allowed utilities to count such savings toward mandated efficiency goals.[13] Absent such state pol-

---

12. https://opower.com/results

13. There still remain issues of how public utilities commissions in their regulatory role should accurately measure and validate behavior change programs, particularly those meant to motivate long-term behavioral changes.

icies and rule changes, utilities had no incentive to pay Opower or its competitors to provide the reports.

Changed Energy-Efficiency Regulations

For the residential and commercial sector, regulatory mandates; policies and programs for appliances, lights, and heating/cooling systems; and other energy-using devices have reduced energy use. Automobile fuel-efficiency standards—the CAFE standards—have been described above. Here I turn to non-transportation energy-efficiency standards.

Federal appliance efficiency standards were first legislated as a direct response to the energy crisis of 1973–74, as part of the Energy Policy and Conservation Act (EPCA) of 1975; they were developed under the Ford administration and signed into law by President Gerald Ford.[14] The 1978 National Energy Conservation Policy Act (NECPA), signed by President Jimmy Carter, introduced manufacturer impact analysis to the rule-making process. The regulatory authority was expanded by the National Appliance Energy Conservation Act of 1987, under Ronald Reagan; the Energy Policy Act of 1992, under George H. W. Bush; and the Energy Policy Act of 2005, under George W. Bush. These regulations were truly bipartisan.

Federal lighting standards were enacted in the Energy Independence and Security Act of 2007 and signed by President George W. Bush. Under this law, screw-based lightbulbs must use at least 27 percent less energy by 2014, and most lightbulbs must be 60–70 percent more efficient than the standard incandescent lights by 2020.

---

14. Taylor, Spurlock, and Yang (2015) provide more detail on the history of federal minimum efficiency performance standards (MEPS). Although the 1975 EPCA called for federal MEPS to be implemented by 1980, for a variety of reasons they were delayed, so the first standards did not come into effect until 1987. But California's first MEPS came into effect in 1977. Although Federal MEPS were designed to preempt, or supersede, state standards, a series of delays, legal agreements, waiver procedures, and more have mediated this for different product standards.

A wide variety of policy interventions, described well by the National Research Council, have been used to enhance private investments in energy efficiency:

> Formidable market barriers . . . tend to block the development and introduction of energy-efficient technologies in the buildings sector even if paybacks in the form of reduced energy bills are very rapid. Over the 20-year horizon of this study, a potent response has emerged that marries federal R&D-based technology innovation initially with financial incentives for technology adoption (typically funded by utility programs and tax incentives) and ultimately with amendments to building and equipment efficiency standards.[15]

The formidable barriers were those outlined in Chapter 1 (see Table 1.1. Some Barriers to Energy-Use Optimality).

The system of policy interventions discussed in the National Research Council Report is illustrated conceptually by a diagram (Figure 6.7) taken from presentations given by ENERGY STAR appliance program officials.[16] Figure 6.7 illustrates the idea that (1) R&D creates options for substantially increasing the highest levels of energy efficiency of particular products; (2) state or federal building codes and standards set lower limits on the energy efficiency of products that can be sold; and (3) ENERGY STAR labels those products with superior energy efficiency in order to encourage their purchase.

Two important forces need to be added to Figure 6.7 to more completely describe the system of policy interventions. Once R&D (funded by either the private or public sector) creates options, private-sector

15. *Energy Research at DOE, Was It Worth It? Energy Efficiency and Fossil Energy Research 1978 to 2000*, Committee on Benefits of DOE R&D on Energy Efficiency and Fossil Energy, Board on Energy and Environmental Systems, Division on Engineering and Physical Sciences, National Research Council, p. 27.

16. R. Karney, "ENERGY STAR Criteria for Clothes Washers: Overview of ENERGY STAR Criteria Setting Process and History of Clothes Washer Criteria" (2004). The diagram appearing in this book was copied from Taylor, Spurlock, and Yang (2015) at http://eetd.lbl.gov /publications/confronting-regulatory-cost-and-quali.

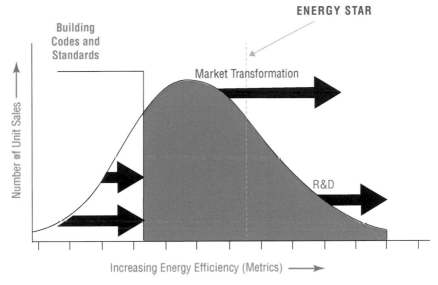

Figure 6.7. The Conceptual Interplay of Governmental Regulatory Policy Instruments

Source: Taylor Spurlock Yang, R_Rvw_MEPS_for_RFF - 20151020_combined-1.pdf, p. 24

firms develop, produce, and market products that offer these high levels. In some states, utilities and state energy policy push and pull the technology, through upfront rebates, information, and even financing, to speed diffusion of the new products.

These various forces, working together, increase the fraction of sales of more energy-efficient products and decrease the fraction of sales of less-efficient products. When successful, this combination of market pushes and pulls over time leads to a market transformation for those products, resulting in either mandatory use of the products through building codes and standards or general uptake in the private marketplace.[17]

---

17. In 2007, California adopted a similar approach directing development of a strategic plan (formally, the 2008 California Long-Term Energy Efficiency Strategic Plan) that would articulate how energy-efficiency programs in California will be designed with the goal of transitioning to either the marketplace without ratepayer subsidies, or to codes and standards. CA Public Utilities

Refrigerators' efficiency gains, described in Chapter 2, provide an excellent example of this pattern. Energy use data trends were presented in Figure 2.2 (New Refrigerator Energy Use, Volume, and Price Trends). As a direct response to the increasing use of electricity in refrigerators, in 1977 the DOE, through Oak Ridge National Laboratory, supported research into compressor efficiency, with an intention to decrease electricity use by refrigerator-freezers and supermarket refrigeration systems. "DOE targeted both improved components, starting with the electricity-intensive refrigerator compressor, and computer tools for analyzing refrigerator design options. Early successes included a compressor system that achieved 44 percent efficiency improvement over the technology commonly used in refrigerators of the late 1970s."[18]

An innovative incentive was created for refrigerator manufacturers to conduct R&D targeted toward energy efficiency—the Super Efficient Refrigerator Program (SERP). The first "Golden Carrot" program to be implemented in the US, SERP was a competition among refrigerator manufacturers designed to accelerate development and commercialization of super-efficient refrigerators. SERP featured a $30 million award, competitively given to the refrigerator manufacturer that could develop, distribute, promote, and sell the most energy-efficient, CFC-free refrigerator/freezer in the most cost-effective manner possible.[19] Whirlpool was the ultimate winner, and the SERP led

---

Commission, Decision 07-10-032. The 2008 Strategic Plan (p. 5) identified five policy tools that can be used to "push" or "pull" more efficient products or practices to market: customer incentives, codes and standards, education and information, technical assistance, and RD&D of emerging technologies.

18. *Energy Research at DOE, Was It Worth It? Energy Efficiency and Fossil Energy Research 1978 to 2000*, Committee on Benefits of DOE R&D on Energy Efficiency and Fossil Energy, Board on Energy and Environmental Systems, Division on Engineering and Physical Sciences, National Research Council, p. 95.

19. With consent of public utilities commissions, six utilities formed SERP Inc. for the Golden Carrot program, and ultimately 18 other public and private utilities, in partnership with the US EPA and environmental groups, joined the effort. The utilities promised to provide rebates in their service areas, thereby increasing the market share for the winner.

to other efforts to combine federal R&D efforts with upstream manufacturing incentives.[20] SERP apparently increased the amount of energy-efficiency R&D conducted by Whirlpool and competing refrigerator manufacturers.

Once economically attractive reductions in energy use were possible, refrigerator manufacturers embedded this technology into some models they put on the market. Utilities, particularly those part of SERP, had promised to provide rebates in their service areas, thereby providing a market pull to increase market share for the winner.

Once energy-efficient refrigerators were successfully commercialized, the State of California imposed a sequence of appliance efficiency standards for refrigerators sold in California. Additional private-sector and public-sector research was further motivated by these standards, and this research led to additional efficiency improvements.

In turn, those additional efficiency improvements subsequently led to even tighter California efficiency standards. Federal standards followed, each requiring lower electricity use than the previous standards. As a result of this combination of private-sector and public-sector actions, the average electricity consumption of a new refrigerator—about 1800 kWhr per year in 1972—was decreased to below 500 kilowatt hours per year in 2014. Figure 2.2 (New Refrigerator Energy Use, Volume, and Price Trends), shown in Chapter 2, includes data of when standards were promulgated by the State of California or by the federal government, although it does not show the times of the technology advances.

In addition, as described above, the ENERGY STAR voluntary program successfully encouraged manufacturers to market refrigerators that qualify for the ENERGY STAR label in addition to those that just

---

20. See "The Super Efficient Refrigerator Program: Case Study of a Golden Carrot Program," NREL/TP-461-7281, (July 1985), http://www.nrel.gov/docs/legosti/old/7281.pdf.

meet the minimum regulatory requirement. And ENERGY STAR has encouraged a significant fraction of buyers to purchase refrigerators that carried that label.

It was this combination of factors—technological advances funded by the US DOE, advances by private-sector refrigerator manufacturers aided by funding from utilities through state public utility commission-approved programs, appliance efficiency standards which progressively pushed and pulled newer technologies into the marketplace, and ENERGY STAR labeling—that reversed the trend of annual increases in energy use by new refrigerators and led to the increases in energy efficiency. The net result has been a fundamental transformation of the refrigerator market.

The federal government has relied heavily on this combination of policies for appliances, partially under the belief that the purchaser of an appliance generally would have little or no knowledge of its future energy costs. With little or no knowledge, manufacturers would put too little attention into improving the energy efficiency of the appliances they offer for sale.

Table 6.1 summarizes the federal equipment efficiency standards for appliances typically sold in the residential sector. There is an additional list of appliances subject to equipment efficiency standards in the commercial/industrial sector.

Table 6.1 also shows that for some of these products, in addition to refrigerators, there are state regulations, many more stringent than the federal standards. California, Oregon, and Connecticut are particularly active appliance-efficiency regulators. Although there are no federal standards for televisions, compact audio equipment, DVD players, or portable electric spas, all three states have set standards for these devices and altered the marketplace even absent federal action.

California's television appliance efficiency standards provide one example. As of 2006, the maximum standby electric load for televisions

PRODUCT COVERED	INITIAL LEGISLATION	LAST STANDARD ISSUED	EFFECTIVE DATE	ISSUED BY	UPDATED DOE STANDARD EXPECTED	POTENTIAL EFFECTIVE DATE	STATES WITH STANDARD
Battery Chargers	EPACT 2005	None	None	N/A	2015	2017	CA, OR
Boilers	NAECA 1987	2007	2012	Congress	2016	2021	
Central Air Conditioners and Heat Pumps	NAECA 1987	2011	2015	DOE	2017	2022	
Clothes Dryers	NAECA 1987	2011	2015	DOE	2017	2021	
Compact Audio Equipment	None						CA, OR, CT
Computers and Battery Backup Systems	None		N/A				
Dehumidifiers	EPACT 2005	2007	2012	Congress	2016	2019	
Direct Heating Equipment	NAECA 1987	2010	2013	DOE	2016	2021	
Dishwashers	NAECA 1987	2012	2013	DOE	2015	2019	
DVD Players and Recorders	None						CA, OR, CT
External Power Supplies	EPACT 2005	2014	2016	DOE	2015	2017	CA
Furnace Fans	EPACT 2005	2014	2019	DOE	2020	2025	
Furnaces	NAECA 1987	2007	2015	DOE	2016	2021	

Table 6.1. Federal Equipment Efficiency Standards: Residential

Source: www.appliance-standards.org/

PRODUCT COVERED	INITIAL LEGISLATION	LAST STANDARD ISSUED	EFFECTIVE DATE	ISSUED BY	UPDATED DOE STANDARD EXPECTED	POTENTIAL EFFECTIVE DATE	STATES WITH STANDARD
Game Consoles	None		N/A				
Microwave Ovens	NAECA 1987	2013	2016	DOE	2019	2022	
Miscellaneous Refrigeration Products	None			DOE	2016	2019	CA
Pool Heaters	NAECA 1987	2010	2013	DOE	2016	2021	
Pool Pumps	None						AZ, WA, CA, CT
Portable Electric Spas	None						AZ, OR, WA, CA, CT
Ranges and Ovens	NAECA 1987	2009	2012	DOE	2015	2018	
Refrigerators and Freezers	NAECA 1987	2011	2014	DOE	2018	2021	
Room Air Conditioners	NAECA 1987	2011	2014	DOE	2017	2020	
Set-top Boxes	None			N/A			
Televisions	NAECA 1987	None	None	N/A	None	None	CA, CT, OR
Water Heaters	NAECA 1987	2010	2015	DOE	2016	2021	

Table 6.1. (continued)

sold in California was 3 watts; as of 2011 that standard had been reduced to 1 watt for all but the larger screen sizes. This effort was part of a broader campaign in California to reduce electricity use by devices that were turned off. As early as 2006, California had set standards for passive use of energy, particularly for DVD players and televisions in their standby mode, not simply for the times that they are operating.

In addition, there are state-specific whole-building standards for new construction. In California, for example, the Title 24 building standards—which were first enacted in 1978—continue to ratchet down the energy use that would be expected in newly constructed buildings. California has set a *goal* that its standards will require that most homes built in 2020 and thereafter be "zero net energy" and that commercial buildings do likewise by 2030,[21] thereby requiring solar energy systems (or other such distributed energy) in addition to energy efficiency.

The California standards for appliances have had impacts well beyond California. California is a large enough market that manufacturers want to sell into that state. But it is costly to design consumer products that have lower electricity consumption for California sales and higher for non-California sales. Therefore, California standards have tended to become de facto standards for the rest of the United States.

An important question is whether the federal and state standards do in fact lead to economically efficient reductions in energy use or whether they are so stringent that they are economically inefficient.[22]

---

21. See 2008 California Long-Term Energy Efficiency Strategic Plan, discussed above, p. 6.

22. For example, it is not clear that California's goal of "zero net energy" for all homes would be economically efficient. For many homes, particularly in shaded locations, solar energy systems are likely to continue being costlier than the value of electricity they generate. These systems would not be economically efficient. However, it is also not clear that the California regulatory system will actually reach this goal.

This issue has been examined recently for federal standards, and the evidence is that the regulations have indeed promoted economic efficiency.

The federal statutory language on setting mandatory efficiency standards is consistent with the idea that regulations should be set at economically efficient levels. The statutory language underlying federal efficiency standards states that they should be set at "the maximum percentage improvement" that is "technologically feasible and economically justified." Similar concepts underlie California regulations set by the California Energy Commission.

Of course, there can be a difference between the intent or language of laws and the actual results. But evidence at the federal level supports the conclusion that the actual results are economically justified. A recent retrospective review by Taylor, Spurlock, and Yang of the cost, quality, and energy use of five appliances regulated by federal efficiency standards since 1987—room air conditioners, refrigerator-freezers, dishwashers, clothes washers, and clothes dryers—indicates that, at least for these five products, the federal rule-making process was effective in choosing economically appropriate stringency of product standards, assuring that the energy savings were economically efficient.[23] That study concluded that if anything, federal regulatory analysts were overly cautious in setting standards. In each case, the analysts overestimated the resulting price of regulatory-compliant products and underestimated their energy performance. In addition to meeting the cost-effectiveness goals, other dimensions of product quality (e.g., capacity, functionality, noise, etc.) tended to improve for all five products.

Although this study examined only five major classes of appliances subject to federal regulations, it does suggest that the efficiency stan-

---

23. See Taylor, Spurlock, and Yang, 2015, at http://eetd.lbl.gov/publications/confronting -regulatory-cost-and-quali.

dards have been consistent with the definition of energy efficiency used in this book—economically efficient reductions of energy use.

Utility Customer-Funded Programs

Regulation of electric and natural gas utilities has also changed in many states, with utility commissions adopting policies implementing "decoupling" mechanisms, which essentially decouple a utility's profits or net revenues from its sales of electricity.

Under traditional utility rate making, a utility has an incentive to resist energy efficiency: its regulated structure for selling electricity or natural gas includes a substantial amount of fixed cost recovery in its rates. A successful energy-efficiency program, absent decoupling, would result in the utility not recovering all of these fixed costs and thus losing money. With decoupling, the state regulatory commission implements regular automatic annual rate adjustments which compensate for under- or over-collection of fixed costs during the previous year. Decoupling eliminates the incentive for utilities to increase their sales of natural gas and electricity and eliminates the incentive against utilities working with their customers to adopt energy-efficient changes.

More than half the states have adopted decoupling mechanisms for either electric or natural gas utilities, as shown in Figure 6.8.

Decoupling alone does not create an incentive for energy-efficiency initiatives run by utilities; it merely eliminates disincentives. Some states have created such incentives through the authority of public utilities commissions to set rates for electric and gas utilities. For example, California has created targets for energy-efficiency programs run by utilities and allows utilities to increase their profits based on the energy-efficiency gains of their customers.

Many state utility commissions (or state legislatures) have set energy-efficiency targets for the gas and electric utilities in their states.

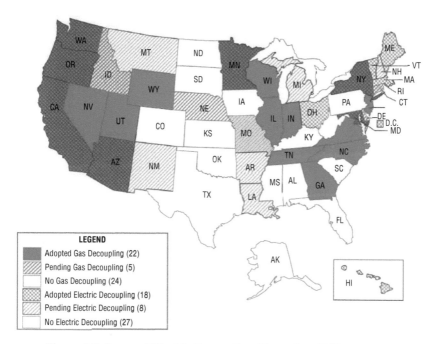

Figure 6.8. Gas and Electric Decoupling: November 2015

Source: Natural Resources Defense Council

These states generally allow the utilities to recover costs of programs, including rebates, marketing materials, and financing programs to customers, through small charges on distribution rates. Most states require that the programs be "cost-effective"—that the overall cost savings from the programs to the utility energy system be greater than the costs of the program.

In California, for example, these programs have led to net energy savings[24] reported by the California utilities of about 6400 gigawatt

---

24. Net savings measure changes in electricity consumption specifically attributable to the energy-efficiency program, taking into account participant behavior.

hours (GWhr) over the years 2010 through 2012, and about 2400 net GWhr over the years 2013 and 2014, about half of 1 percent of electricity consumed during 2010 through 2014. Other states, such as Vermont and Hawaii, have placed responsibility for administering utility–customer funded programs with third-party, non-utility administrators, who have no built-in conflict such as the utilities over reducing sales by efficiency efforts.

The majority of the utility measures covered to date have been in lighting and for compact fluorescent lights in particular, although most programs are now moving to LEDs. In California, subsidized CFLs were often sold for less than 50 percent of the unsubsidized price and at some times were given away in limited quantities. These programs have been successful in accelerating the rate of adoption of energy-efficient lighting, despite the poor quality of CFLs. For example, Figure 6.9 shows the data reported by the California investor-owned utilities to the California Public Utilities Commission for net savings (not necessarily validated) for 2013 and 2014.[25] This shows the heavy concentration on lighting programs, although that concentration is much less than in the previous three years.

State utility programs may also include other measures; the widely used "DEER" database[26] includes dozens of measures and technologies relied upon by utilities and others in their energy programs. Examples include rebates to help replace old refrigerators with ENERGY STAR models, or small payments offered by utilities to their residential customers for allowing old refrigerators to be retired from service.

---

25. http://eestats.cpuc.ca.gov/Views/EEDataPortal.aspx. A recent article by Dian Grueneich, Senior Research Scholar at PEEC, discusses a version of Figure 6.9, based on data from 2010–12, and describes the need to diversify efficiency measures sources, as well as track persistency, integrate efficiency efforts with carbon reduction frameworks, and value energy efficiency as part of an evolving electricity grid. D. Grueneich, "The Next Level of Energy Efficiency: The Five Challenges Ahead," *The Electricity Journal* 28, no. 7 (August/September 2015): 44–56.
26. This is the Database for Energy Efficient Resources, http://www.energy.ca.gov/deer/.

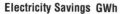

Electricity Savings GWh

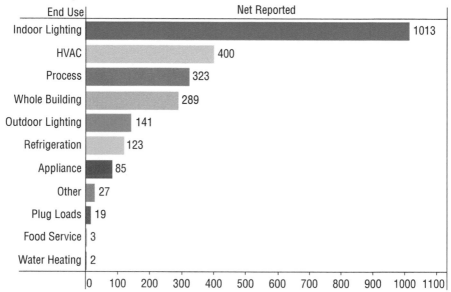

Figure 6.9. California Electricity Savings: Four Largest Investor-Owned Utilities, Two-Year Total, 2013–2014

Source: http://eestats.cpuc.ca.gov/Views/EEDataPortal.aspx

Financial Incentives

In addition to the utility-based, state-funded efforts discussed above, federal and state governments also provide financial incentives, typically in the form of tax credits designed to encourage energy-efficient buildings and technologies. At the federal level, tax credits have been established for energy-efficiency improvement to existing homes, limited to building envelope improvements and heating, cooling, and water-heating equipment. Owners of existing homes can receive a tax credit worth 10 percent of the cost, with an initial limit of $500 in credits.

The federal credits were initially established for a limited time but have been extended many times for one or two years at a time. In 2005, the Energy Policy Act first established tax credits for 2006 and 2007. The Energy Improvement and Extension Act of 2008 extended the credit to 2009. The American Recovery and Reinvestment Act of 2009 added the year 2010 and changed the cap to $1500. The credit option has been extended several times, although the limit was reduced to its original cap of $500. The most recent extension was the Tax Increase Prevention Act of 2014, which extended tax credits for some energy-efficiency measures through December 31, 2014. Only a very limited set of options still remain, primarily options to generate electricity in the residence.[27]

For commercial building owners, the 2005 Energy Policy Act allows a tax deduction of up to $1.80 per square foot for commercial buildings that reduce energy through interior lighting, heating and cooling, and building envelope.

Some state governments, such as California and New York, have added state tax incentives for energy-efficiency investments. These credits are in addition to the federal tax credits and thus are added to the federal incentives.

The tax incentives reduce the capital cost needed by individuals and businesses to make energy-efficiency upgrades, and therefore encourage investments. Although they generally are relatively small fractions of the capital costs, these tax credits have been shown to add an additional behavioral impact and motivate investments, particularly if coupled with other incentives. By bringing to the attention of taxpayers that there are options for gaining tax credits, these credits have further encouraged people to consider whether there is a financial benefit for energy-efficiency investments. This informational

---

27. Still remaining are only geothermal heat pumps, solar energy systems, small wind turbines, and residential fuel cells and microturbines. These credits are currently scheduled to expire December 31, 2016.

nudge probably has had a significant impact in addition to the actual financial subsidy.

In addition to the tax incentives, there are some direct payment programs for buildings and equipment. For example, the US DOE manages a Weatherization Assistance Program that provides grants to states, territories, and some Indian tribes to deeply subsidize or completely pay for energy-efficiency measures for low-income families. The Weatherization Assistance Program is applicable to existing residential and multifamily housing with low-income residents. The program includes measures for the building envelope, its heating and cooling systems, its electrical system, and some electricity-consuming appliances. This program appears to be partially an energy-efficiency program and partially a welfare program that reduces energy costs and increases comfort and health[28] for low-income people, and it may at times be used as an employment stimulus program.[29] Typically demand for the funds far outstrips the actual funding. A number of states, such as California, New York, and Massachusetts, provide additional funding and provision of measures at no cost to low-income utility customers.

Financial incentives have also been established for light-duty vehicles based on emerging technologies. The Energy Policy Act of 2005

---

28. Low-income families with poorly insulated buildings may save money by limiting their home heating. With energy-efficiency upgrades, these individuals may find it in their interest to increase heating up to a more conventional level, therefore providing additional benefits for these families while increasing the use of energy.

29. A recent study by Meredith Fowlie, Michael Greenstone, and Catherine Wolfram argues that this particular program, at least for their sample in Michigan, is not economically efficient: "The findings suggest that the upfront investment costs are about twice the actual energy savings. Further, the model-projected savings are roughly 2.5 times the actual savings." There have been many criticisms of that study, but that discussion is too extensive for this review. M. Fowlie, M. Greenstone, and C. Wolfram, *Are The Non-Monetary Costs of Energy Efficiency Investments Large? Understanding Low Take-up of a Free Energy-Efficiency Program*, E2e Working Paper 016, joint initiative of the Energy Institute at Haas at the University of California, Berkeley; the Center for Energy and Environmental Policy Research (CEEPR) at the Massachusetts Institute of Technology; and the Energy Policy Institute at Chicago, University of Chicago.

established a temporary tax credit for purchases between 2006 and 2010 of some energy-efficient vehicles, particularly hybrid vehicles. In 2005, the credits ranged from $650 for a Honda Accord Hybrid to $3150 for the Toyota Prius. By 2008, most of the credits had been phased out.

Subsequent to 2005, Congress established tax credits for plug-in electric vehicles. Specifically, the Energy Improvement and Extension Act of 2008 and the American Clean Energy and Security Act of 2009 grant tax credits for new qualified plug-in electric drive motor vehicles. This act also provides subsidies for battery electric vehicles.

A special type of financial incentive is a competitive award given to an actual or potential manufacturer of a product. In the energy context, these are often referred to as "Golden Carrot" programs. Golden Carrot programs are created to accelerate the development and commercialization of energy-efficient products. Discussed earlier, the Super Efficient Refrigerator Program was the first Golden Carrot program in the United States. It offered a $30 million agreement competitively awarded to the refrigerator manufacturer who could create and sell a very energy-efficient, CFC-free refrigerator/freezer.[30] Such competitions provide to the winner not only a financial award, but recognition that it can use in marketing its products.

Finally, an emerging financial incentive in wholesale electricity markets is designed to utilize energy efficiency to reduce the time variability of electricity consumption. Energy efficiency is now compensated in a few geographically extensive forward (capacity) wholesale markets (PJM, ISO-NE, and to a lesser extent, MISO). It is likely that other wholesale electricity markets will examine ways of further integrating energy efficiency, at least to reduce the electricity consumption peaks.

---

30. See Jan B. Eckert, *The Super Efficient Refrigerator Program: Case Study of a Golden Carrot Program*, National Renewable Energy Lab, 1995.

Energy Research and Development

The remarkable historical reduction in energy intensity of the US economy has depended upon many different factors. One factor has been ubiquitous throughout the changes: technology innovations and innovations in energy management and utilization. Many of these innovations were direct or indirect results of research and development conducted by universities, research institutes, national labs, governmental agencies, and private-sector firms.

The majority of the dollars spent on energy-related research and development has been spent by the private sector, but the federal government has also played a very important role in working with universities, national labs, industry in energy R&D, and in stimulating private-sector R&D. This was described by the National Research Council report:

> From 1978 through 1999, the federal government budgeted $91.5 billion (2000 dollars) in energy R&D, mostly through DOE programs (NSF, 2000). This direct federal investment constituted about a third of the nation's total expenditure on energy R&D, the balance having been spent by the private sector. Since government policies— from cost sharing to environmental regulation to tax incentives— influenced the priorities of a significant fraction of the private investment, it can be said that, on balance, the government has been the largest single source and stimulus of energy R&D funding for more than 20 years.[31]

The National Research Council report also provided its assessments of some of the most important fossil energy and energy-efficient

---

31. *Energy Research at DOE, Was It Worth It? Energy Efficiency and Fossil Energy Research 1978 to 2000*, Committee on Benefits of DOE R&D on Energy Efficiency and Fossil Energy, Board on Energy and Environmental Systems, Division on Engineering and Physical Sciences, National Research Council, 2001, p. 9.

Technology Now in the Marketplace	Level of DOE Influence
Energy efficiency	
More efficient electric motors	A/M
Higher mileage automobiles	A/M
More efficient electronic ballasts	D
More efficient household refrigerators	D
More effective insulation	I
Synthetic lubricants	A/M
More efficient gas furnaces	A/M
More energy-efficient windows	I
More efficient industrial processes	A/M
More efficient buildings	I

NOTE: Influence levels: A/M, absent or minimal; I, influential; D, dominant.

Figure 6.10. The Most Important Fossil-Energy and Energy-Efficiency Technological Innovations since 1978

Source: The National Research Council report

technology innovations since 1978. This table from the report,[32] shown here as Figure 6.10, includes these energy-efficiency innovations; the portion of the table on fossil energy innovations has been deleted. Federal government influence has been dominant for some of the technologies, influential for others, and for some minimal at best. The National Research Council study supports the conclusion that the federal government, in particular the DOE, has been very important in a large fraction of the most important energy-efficiency technology innovations.

The policy case for R&D in energy rests on the observation that even though public benefits exist (e.g., national security improvement and environmental protection), individual companies cannot capture these benefits. This barrier has been referred to as "R&E Spillovers" and

---

32. Ibid., p. 13.

"Externalities" in Table 1.1 (Some Barriers to Energy-Use Optimality). Energy R&D, particularly related to the use of energy, therefore can lead to significant benefits, benefits that might not be captured by the private sector or that may be captured only with a substantial time lag.

The National Research Council study—*Energy Research at DOE, Was It Worth It? Energy Efficiency and Fossil Energy Research 1978 to 2000*[33]—in 2001 examined the costs and benefits of applied R&D at the DOE, in order to examine whether public benefits exceeded the public costs and, if so, by how much.[34] That study unequivocally concluded that the benefits exceed the cost by a substantial amount[35] and that the R&D was a good investment of public funds.

> The committee found that DOE's RD&D programs in fossil energy and energy efficiency have yielded significant benefits (economic, environmental, and national security-related), important technological options for potential application in a different (but possible) economic, political, and/or environmental setting, and important additions to the stock of engineering and scientific knowledge in a number of fields.

Although federal energy R&D has been beneficial for all three energy-policy-triangle goals, the federal government has not maintained high funding for this R&D. In the two decades before the 1973–74 energy crisis, the US federal government had allocated relatively little nondefense funds to energy R&D. During the decade after the energy crisis, federal energy R&D funds increased sharply, both in absolute terms and as a fraction of nondefense R&D. However, once

---

33. *Energy Research at DOE, Was It Worth It? Energy Efficiency and Fossil Energy Research 1978 to 2000*. Committee on Benefits of DOE R&D on Energy Efficiency and Fossil Energy, Board on Energy and Environmental Systems, Division on Engineering and Physical Sciences, National Research Council, 2001, p. 104.

34. James Sweeney was one of the members of this committee and was involved in writing some of the conclusions repeated here.

35. Ibid. p. 5.

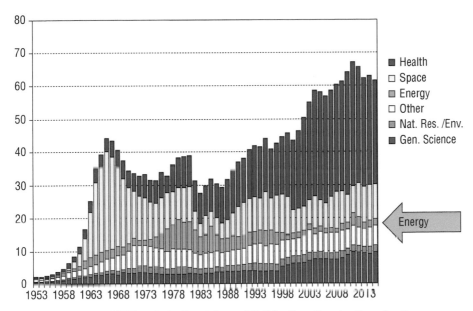

Figure 6.11. US Trends in Nondefense R&D by Function (outlays for the conduct of R&D, billions of constant FY 2014 dollars)

Source: AAAS, based on OMB Historical Tables in Budget of the United States Government. FY 2015 is the President's request. Some Energy programs shifted to General Science beginning in FY 1998.

the crude oil prices dropped precipitously in 1986, federal nondefense energy R&D began declining. In recent years, the federal nondefense energy R&D expenditures declined to levels roughly equivalent to the pre-energy-crisis levels in absolute terms. As a fraction of the R&D budget, energy R&D is now even smaller than it was before the energy crisis. This amount of funding is not sufficient to cover the important energy needs. Unfortunately, this large reduction in energy R&D can be expected to decrease the rate of innovation, including innovation designed to reduce energy use. The expenditure data are shown in Figure 6.11.

Not all funding for R&D is based on federal funds. Companies also conduct internally funded energy research. For example, firms selling energy-using devices such as automobiles, airplanes, or lights have had incentives to conduct competitive research. Universities and other research organizations have continued or expanded research into energy issues, funded in some cases by the federal government, in some cases by foundations or companies, and in other cases by individuals.[36] California and New York State have had a large energy R&D program for decades. These various non-federal funding sources appear to remain vibrant even with the reduction of federally funded energy R&D. However, the case remains strong for additional federal funding of energy R&D.

Energy Policy and Advocacy Organizations

As indicated in Chapter 3, Congress has created several federal agencies charged with addressing energy problems. The Federal Energy Administration (FEA) and the Energy Research and Development Administration (ERDA) were created in 1974. Subsequently, in 1977, these two agencies became part of the newly created Department of Energy. Also in 1974, the International Energy Agency (IEA) was founded as a Paris-based intergovernmental organization under the Organisation for Economic Co-operation and Development (OECD).

State governments followed not long afterward. For example, the California Energy Commission (CEC) was established in 1974 by the state Legislature as California's primary energy policy and planning agency. Its responsibilities include "[p]romoting energy efficiency and conservation by setting the state's appliance and building energy efficiency standards."[37] Founded in 1975, the New York State Energy Research and Development Authority (NYSERDA), "promotes energy

---

36. For example, the Stanford Precourt Energy Efficiency Center and the Precourt Institute for Energy owe their existence to and are funded primarily by gifts from a private individual, Jay Precourt. Stanford's Global Climate and Energy Project has been funded entirely by commitments from several private-sector companies. The initial companies contributed $225 million: ExxonMobil, General Electric, Toyota, and Schlumberger.
37. http://www.energy.ca.gov/commission/

efficiency and the use of renewable energy sources. . . . Collectively, NYSERDA's efforts aim to reduce greenhouse gas emissions, accelerate economic growth, and reduce customer energy bills."[38] In fact, every state has both a State Energy Office (though many are not as comprehensive or as fully staffed as the CEC and NYSERDA). Each state also has a public utilities commission that regulates, to varying degrees, utilities serving businesses and residences in the state. As discussed below, these commissions play a critical role in the US energy-efficiency landscape.

State regulatory commissions in many states used their authority to change the behavior of utilities, as discussed above. Public utilities commissions have adopted critical policies and allocated billions of funding dollars to engage utilities in energy efficiency, leveraging the ability of utilities to work directly with their customers. In many states utilities themselves became strong advocates for additional energy-efficiency initiatives.

It is not simply governmental organizations that have turned their attention to energy issues. In the last four decades, multiple non-governmental organizations (NGOs) added energy-efficiency activities and new NGOs were created to help address some energy and climate problems. The Natural Resources Defense Council (NRDC) and the Environmental Defense Fund (EDF) added important energy and climate activities in the 1970s, recognizing the very tight link between energy production and local/global environmental impacts. These two organizations have played leading roles in bringing energy-efficiency options to the table and in promoting energy efficiency.

The Alliance to Save Energy was founded in 1977; it "promotes energy efficiency worldwide."[39] The American Council for an Energy-Efficient Economy (ACEEE) was founded in 1980 "as a catalyst to

---

38. http://www.nyserda.ny.gov/About
39. http://www.ase.org/about/history-mission

advance energy-efficiency policies, programs, technologies, investments, and behaviors."[40] The Rocky Mountain Institute, founded by Amory and Hunter Lovins in 1982, has a purpose "to drive the efficient and restorative use of resources,"[41] particularly energy. The Energy Foundation started at a later time—1991—is devoted to changing energy policy to support clean technologies. Founded in 1991, the Appliance Standards Awareness Project "organizes and leads a broad-based coalition effort that works to advance, win and defend new appliance, equipment and lighting standards."[42]

In addition, six regional organizations have been founded to aggressively promote energy efficiency: in the Southeast (SEEA), Midwest (MEEA), Northwest (NEEA), Northeast (NEEP), Southwest (SWEEP), and Texas/Oklahoma (SPEER). All work through funded partnerships with the US DOE, utilities, and other organizations, providing technical assistance to states and municipalities to support efficiency policy development and adoption, program design and implementation.[43]

Sector-specific organizations have also been created. For example, the US Green Building Council, founded in 1993 with a goal "to promote sustainability in the building and construction industry,"[44] is responsible for LEED certifications, as discussed above. CALSTART, founded in 1992 in California, but now operating worldwide, is "dedicated to supporting a growing high-tech, clean transportation industry."[45] Founded in 1991, the Consortium for Energy Efficiency (CEE) is a consortium primarily of natural gas and electricity efficiency program

---

40. http://aceee.org/overview-mission
41. http://www.rmi.org/Vision and Mission
42. http://www.appliance-standards.org/content/mission-and-history
43. http://www.seealliance.org/wp-content/uploads/Resource-REEOEEPolicy2014.pdf
44. http://www.usgbc.org/about/history
45. http://www.calstart.org/About-us/Who-We-Are.aspx

administrators from the United States and Canada, working together to accelerate energy-efficient products and services. The CEE's role is "to influence national players—manufacturers, stakeholders, government agencies—to maximize the impact of efficiency programs."[46] The Green Grid, a consensus-driven consortium to promote resource efficiency in information technology, works to develop globally adopted metrics and measurements to encourage greater resource efficiency in data centers.[47] The Regulatory Assistance Project (RAP), founded in 1992 in the United States and now operating worldwide, has been a leading adviser to state and national policy makers, particularly in electric and gas regulation related to energy-efficiency programs.[48]

These organizations—both governmental and non-governmental— have been important for energy policy making, energy research, energy information, energy advocacy, program development, and implementation of energy efficiency. Each has placed continued emphasis on energy efficiency and each has continued to keep public attention on energy efficiency. Some impacts have been described in the previous sections, in R&D for new technology, in promoting the broad implementation of technology, in encouraging industry energy-efficiency innovations, in labeling, in providing energy information, in advocating for regulatory policies, and in creating industry support for energy-efficiency regulations. These organizations ranged from important to critical in the various energy-efficiency advances described above.

46. http://www.cee1.org/content/about
47. http://www.thegreengrid.org/
48. http://www.raponline.org/about

7

Policy Lessons from the Past Forty Years
WHAT HAS LED TO INCREASED ENERGY EFFICIENCY?

Perhaps the most important conclusion is that the cumulative, broadly distributed growth in energy efficiency has been the result of many factors working together, not simply any one of these factors.

Efforts to attribute changes to just one factor—be it market competition, regulation, higher prices, increased awareness, utility programs, or nudges—miss the important lesson from energy-efficiency history. All of these factors have been mutually reinforcing in energy-efficiency gains. And these gains did not occur in a single place. Rather they have been broadly distributed in companies, government agencies, households, and in the transportation system.

Beginning in 1973–74, the sharp increase in oil prices led to increases in energy prices throughout the economy. These higher prices provided a motivation for change. Many products were improved by reducing the use of energy, sometimes increasing their production cost while reducing operating costs, and sometimes not significantly increasing production cost. This process of innovation often led to other valuable product quality improvements as well. Newly developed products were aimed at reducing energy use. Managerial practices changed so as to motivate managers and other employees to take energy use into account in operations. Companies faced incentives to purchase equipment that was less energy intense.

The heightened awareness of energy problems—insecurity of energy imports and environmental degradation from greenhouse gas emissions—led to the creation of state and federal governmental organizations as well as non-governmental organizations. These various

163

organizations played important roles in keeping attention on energy problems and in particular in keeping attention on energy efficiency as the most effective strategy for dealing with this array of energy problems.

Governments, particularly the federal government, for somewhat over a decade after the energy crisis, increased their investment in R&D in many energy areas, including energy efficiency. Some investment came through national laboratories, universities, or other research organizations; some came through cost sharing with private sector firms; others were encouraged by tax incentives. Energy-efficiency research continued in some firms selling energy-using devices such as automobiles, airplanes, or lights. Universities and other research organizations continued or expanded research into energy issues, funded in some cases by the federal government, in some cases by private foundations or companies, and in other cases by individuals.

In addition to the increase in energy prices, attitudes about energy changed. Prior to 1973 there was very little attention paid to energy use or its consequences. But the 1973 oil embargo and oil price increase was a wakeup call. Within government agencies, corporations, and individual households there was new attention paid to energy. Although in many households the cost of energy was of relatively little salience because it represented a relatively small portion of the household budget, in many households, people began paying attention to energy use. President Jimmy Carter's fireside chat stressed the importance of not wasting energy. For some people this made a difference in actions while for others it only raised awareness. But increased awareness itself is important for taking energy into account in private decision-making. Non-governmental organizations have successfully kept energy efficiency as part of the national dialogue. State policy efforts have assisted in funding education about benefits of efficiency and in marketing to the general public and businesses.

The federal government and several state governments began promulgating regulations designed to reduce energy use in the 1970s. At the federal level, CAFE standards for cars and light trucks led to a halving of energy use per mile driven. Appliance efficiency standards were created. States strengthened their building standards. California created its own standards for appliance efficiency and buildings, including standards for passive use of energy, for energy used for appliances that were turned off at the time. Lighting efficiency standards were introduced. And non-governmental organizations continued to work with governmental agencies to promote strong regulations.

Subsidies to assist in lowering initial costs, changed the rate and depth of implementation of some of the energy-efficient technologies. Federal tax rules, which allowed deductions of energy-efficiency investments, encouraged building retrofits and increased the rate of market penetration of energy-efficient light-duty vehicles. Utility programs promoted many measures, particularly energy-efficient lighting.

The combination of all of these forces—not one of these forces acting alone—has been responsible for the remarkable increases in energy efficiency in the US economy. In turn, these energy-efficiency increases have been important for the three goal areas illustrated by the energy policy triangle—security, environment, and the economy. Energy efficiency and new domestic energy supplies have kept net energy imports under control, but energy efficiency has had by far the largest impact. Energy efficiency (and to a small extent new domestic clean energy supplies) have led to a 61 percent decrease in the carbon intensity of the US economy since 1973.

Going Forward: The President's Goal

President Obama, in his 2013 State of the Union address described an "all-of-the-above" approach for further energy progress. Although most

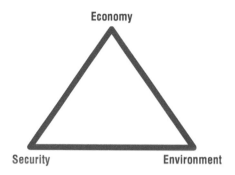

Figure 7.1. The Energy Policy Triangle

of the "all-of-the-above" approach focused on energy supply, he did set a goal for future energy productivity (the mathematical inverse of energy efficiency):

> a new goal to double American energy productivity by 2030 . . . More specifically, the Administration will take action aimed at doubling the economic output per unit of energy consumed in the United States by 2030, relative to 2010 levels.[1]

The president set the goal, but is the goal worth meeting?

This book started with the energy policy triangle (Figure 7.1). Energy policy typically strives for improvements in the health and growth of the economy, protection of the domestic and international environment, and enhancement of domestic and international security. Chapter 1 asserted that energy efficiency is good for all three elements of the energy policy triangle, pointing out that the cleanest energy is the energy you don't need in the first place.

Chapter 2 showed some of the many ways that energy efficiency has been good for the economy by reducing costs for air and ground mobility; lighting homes and businesses; computing; industrial processes;

---

1. https://www.whitehouse.gov/the-press-office/2013/03/15/fact-sheet-president-obama-s-blueprint-clean-and-secure-energy-future

refrigeration; and keeping buildings at comfortable temperatures. Energy efficiency has provided savings for the federal government, with beneficial impacts on the federal deficit and balance of trade deficit. It has reduced system costs for utilities, avoiding construction of expensive power plants and transmission lines.

Chapter 3 showed that small, broadly distributed annual changes in energy intensity accumulated over a 40-year period, reducing the energy intensity of the US economy from 1973 by 57 percent. Energy efficiency since the 1973 energy crisis has reduced the use of energy by 80 quadrillion BTU from what it would have been had there not been the cumulative process of energy-efficiency improvements! Since 1973, consumption reductions resulting from energy efficiency have been far greater than the increased domestic supply of all forms of primary energy taken together.

Chapter 4 showed how good energy efficiency has been for the environment. Reductions in energy intensity of the economy, driven primarily by increases in energy efficiency, have been far more important in reducing the carbon intensity of the US economy, and therefore the carbon dioxide emissions of the United States, than all of the clean energy technologies put together.

Chapter 4 also showed that energy efficiency has been good for security, keeping energy imports to a manageable level. Energy efficiency has allowed the United States to reduce energy imports so that the country may soon become self-sufficient in energy, with energy efficiency having a larger role than all of the domestic energy supply increases put together. In addition, there is a link between global climate change and security. Global climate change has the strong potential of causing armed conflicts and of creating generations of environmental refugees, important issues of international security.

And there is good reason to believe that further enhancements of energy efficiency can continue to have further such benefits. So the president's goal is worth meeting.

But is the president's goal attainable under the policy course currently in place? And in the current market environment of low oil prices? If not, is it likely that the remarkable record of energy-efficiency progress over the past 40 years will no longer continue and might possibly be reversed? We turn to those two alternative views in the final section of this book, first addressing the likelihood of meeting President Obama's articulated goal.

Importantly, for this discussion, the reader should internalize the adage attributed to many different people:[2] "It's difficult to make predictions, especially about the future."

Going Forward: Will the President's Goal Be Met?

Is the president's goal attainable under the policy course currently in place and in the current market environment of low oil prices? In summary, it is highly unlikely.

The president's goal of doubling energy productivity requires energy intensity[3] to decrease an average of 3.4 percent per year for 20 years, beginning in the year 2010. But during the Obama Administration, energy intensity decreased on average only 1.5 percent per year, far short of the president's stated goal. From 2010 through 2014, the rate of energy intensity decline has indeed increased to 1.8 percent per year, faster than during the previous two years, but well below the rate needed for the goal of doubling energy productivity. Given this lower rate of improvement so far, in order to meet the president's announced goal, the energy intensity will have to decrease an average of 3.8 percent per year from 2014 until 2030.

---

2. I like to think that the adage came from Yogi Berra. But variants can be traced to K. K. Steincke, Jesse Markham, Bradford Hill, Neils Bohr, Mark Twain and Yogi Berra. See http://quoteinvestigator.com/2013/10/20/no-predict/.

3. As discussed previously, energy productivity is the mathematical inverse of energy efficiency.

The maximum extended rate of energy intensity decrease in the US economy was 2.7 percent per year from 1973 through 1985, a period of intense energy policy attention, many energy-efficiency policies, and high energy prices. Meeting the president's goal, therefore, is likely to require the United States to go well beyond the conditions that allowed it to maintain the energy-efficiency growth over the last 40 years. These conditions, briefly reviewed in the following paragraphs, need to be supported and magnified to meet the goal of halving energy intensity (doubling energy productivity).

However, although President Obama set this goal, there has been very little federal movement in support of the goal, with the exception of the increased fuel efficiency standards for cars and trucks.[4] Sustained national leadership by the president and Congress—over the rest of the Obama presidency and over the next three presidential terms— would be necessary for success. But at this point, no one knows who will be sworn in as president on January 20, 2017; what actions he or she may want to take; and whether Congress will agree to those actions.

In addition, the federal government has neither the authority nor the capability to achieve the goal without widespread support from many stakeholders in the public and private sector. Active leadership and/or participation by state and local governments, businesses, not-for-profit organizations, and families is needed if there is to be success. Leadership in Congress and the Administration can play a vital role in securing support by those many stakeholders, but only if Congress and the president choose to play such leadership roles.

Assuming that Congress and future presidents support the goal and that they are prepared to work with the many stakeholders, what could

---

4. I do not include the EPA Clean Power Plan because, although energy efficiency can be included in the state implementation plans, the Plan is primarily directed toward clean supplies of electricity.

be done to provide some possibility of meeting the goal of halving energy intensity of the economy by 2030? The history of energy efficiency gives us some guidance.

The first observation is that most energy-efficient technologies are developed by the private sector. Therefore, continuing to maintain a healthy innovative economy is fundamental. If the United States fails to maintain a healthy innovative economy, there will be virtually no chance of meeting the president's goal.

Increased energy prices are important for motivating individuals, companies, and governments to take actions that increase the rate of energy-efficiency improvements. In going forward, it is important not to shield companies and individuals from increased energy costs. One such important energy cost is the externality cost of greenhouse gas emissions. A carbon price reflective of the damages associated with greenhouse gas emissions, if broadly implemented throughout the economy, would help motivate further improvements in energy efficiency. Such a carbon tax, particularly a revenue-neutral carbon tax, should be a crucial component for meeting the president's goal.[5]

But recently the international crude oil prices have dropped precipitously from previous high levels. This was shown in Figure 3.4 (Crude Oil Nominal Prices and Real Prices, 2015 dollars). In addition, natural gas prices are relatively low in the United States, so absent direct governmental actions to increase energy prices and to adopt a substantial carbon price (or other energy tax), price incentives will be substantially less than during much of the 1973–85 time, a time during which US energy intensity decreased on average 2.7 percent per year.

---

5. A carbon tax can be made revenue neutral by either returning carbon tax revenues to taxpayers or by reducing personal and corporate income taxes. Whatever formula is used, in order to have the right incentives, the return to an individual taxpayer should not depend on the carbon taxes that an individual or company actually pays.

The current low energy prices strengthen the case for implementing a substantial carbon tax as part of the strategy to meet the president's goal.

Because R&D plays a crucial role in development of innovative energy-efficient technologies and practices, it is important that the federal government continue to find ways to encourage such R&D investments. Increasing government funding for energy-efficiency R&D from the recently shrunken levels would be an important starting point and would help to accomplish the president's stated goal. Governmental actions need not be limited to direct funding but can also include appropriate tax incentives. Private–public partnerships could allow a leveraging of scarce funds.

Many barriers still exist, impeding or at least slowing down adoption of economically efficient new technologies and practices. For energy-efficient products aimed primarily at individuals and families rather than at sophisticated companies, typically rapid widespread diffusion has required support through state, federal, and local governmental policies, funding, incentives, or regulations. For changed energy-efficient practices, diffusion may be slow, even to companies, and typically requires information/educational initiatives.

Market failures and structural impediments, as well as behavioral issues, can be further attacked in many ways, with different instruments appropriate for the different failures or impediments. As discussed in previous chapters, many such instruments are currently in place, but there is plenty of room for additional instruments or for strengthening the existing instruments.

For example, information barriers are ubiquitous throughout the energy system. One approach has been and can continue to be the provision of broad dissemination of useful, accurate information about energy options, including good information about cost of operating

172   *James L. Sweeney*

appliances and other energy-using devices. Conferences, workshops, and marketing may work for reaching large businesses but may not be sufficient for individuals and small businesses. Efficiency standards continue to be useful, particularly when the energy use of the product— appliance, vehicle, or building—is difficult for the buyer/renter to observe before the purchase or rental agreement is completed. City, county, or state energy-efficiency policies can take into account local circumstances and can be linked to building permitting.

Externality taxes can be implemented to address important areas of externalities, such as emissions of greenhouse gases. In other situations, subsidies are appropriate, for example, to encourage the more rapid adoption of new technologies, particularly when there are important learning-by-doing effects or when consumers learn of the new technologies by word of mouth.

And additional instruments can be invented to motivate further improvements in energy efficiency. Nudges of many types are working, and more can be designed. Other such instruments might be random rewards for energy-efficient behavior; real-time feedback of appliance-level energy use in buildings; educational programs through schools, churches, and youth organizations; demonstrations of buildings, equipment, and processes; initiation of sustainable mobility systems; and the list goes on.[6] There is no shortage of opportunities to create innovative instruments that will enhance the already impressive record of energy-efficiency improvements.

Federal, state, and local governments, as well as many other players, will need to aggressively embrace such innovative instruments to meet the president's announced goal. In going forward, it is important to re-

6. For example, see some examples at Stanford, funded by the Department of Energy, Advanced Research Projects Agency-Energy (ARPA-E), http://peec.stanford.edu /energybehavior/index.php.

alize that these factors are highly interactive and work together. In a real sense, the whole has been and can continue to be substantially greater than the sum of its parts.

However, at this point it appears highly unlikely that the federal government will in fact embrace such instruments.

Given the current low energy prices and the low probability that the federal government will adopt a wide range of additional policies to promote energy efficiency, it appears that the president's goal for energy efficiency is simply unrealistic.

Going Forward: Will Energy-Efficiency Progress Stop?

But what about the opposite possibility? Is it likely that the remarkable record of energy-efficiency progress over the past forty years will no longer continue and possibly be reversed? I find that possibility also to be very unlikely.[7]

The most obvious change is the dramatic drop in the price of crude oil and thus in the price of the refined petroleum products that are used in the transportation sector and for some heating. Figure 3.4 (Crude Oil Nominal Prices and Real Prices, 2015 dollars) shows that the imported crude oil price dropped from $100 per barrel in July 2014 to $30 per barrel in December 2015, just 17 months later. This is roughly equivalent to the drop from $105 per barrel (2015 dollars) in March

---

7. I am not alone in this conclusion. The most recent DOE EIA *Annual Energy Outlook,* in its reference case (tables released April 2015) projects energy use to grow from 2015 through 2040 at an average rate of 0.3 percent per year, while it projects real GDP to grow at an average rate of 2.4 percent per year, implying an energy-intensity decrease at an average rate of 2.1 percent per year. See http://www.eia.gov/forecasts/aeo/tables_ref.cfm. However, those projections were made before the drop in crude oil price. The EIA low oil price case projects almost the same rates of change. See http://www.eia.gov/forecasts/aeo/pdf/appc.pdf. ExxonMobil's 2016 *The Outlook for Energy: A View to 2040* projects US primary energy use to decline on average 0.1 percent per year until 2040, with real GDP growing at an average rate of 2.4 percent per year, implying energy intensity would decrease at an average rate of 2.5 percent per year. See http://cdn
.exxonmobil.com/~/media/global/files/outlook-for-energy/2016/2016-outlook-for-energy.pdf.

1981 to $24 per barrel in July 1986, but is even more obvious because of the suddenness of the change.

Once the crude oil price dropped in 1986, during the subsequent 18 years, real oil prices fluctuated between $14 per barrel and $59, averaging $31 per barrel (all 2015 dollars), about equal to the current price. So as of today, oil prices are similar to those prevailing during those 18 years.

What can be expected for oil prices in the future? The best estimate comes from the futures prices for oil, traded on the New York Mercantile Exchange. Although futures prices at any time cannot be said to predict the prices that will prevail in the future, if for no other reason than that futures prices change rapidly, it is reasonable to understand futures prices as estimating the middle of the probability distribution of those prices in the future, as seen by sophisticated oil traders. After all, prices in futures markets are based on some sophisticated traders believing it is financially beneficial to purchase a contract for future delivery of oil, paying the futures price, and other sophisticated traders believing it is financially beneficial to sell that same contract.[8]

As of today—February 24, 2016—the crude oil futures price[9] is $40 per barrel for deliveries in January 2017, rising to $50 per barrel for deliveries in December 2023.[10] However, options contracts for crude oil—puts and calls—show that there is a wide band of uncertainty around these prices.

---

8. Buyers and sellers may have different goals for risk management, in addition to different beliefs.

9. Contracts specify delivery at Cushing Oklahoma, but contracts can be settled financially with no actual deliveries of the oil.

10. Data access on February 24, 2016 from http://www.cmegroup.com/trading/energy /crude-oil/light-sweet-crude_quotes_settlements_futures.html.

Thus the best estimate is that the crude oil price will rise slowly over the next few years and will thus be somewhat higher in real dollars than the prices that prevailed during the 1986 through 2004 period.[11]

In that period, as was shown in Figure 3.8 (Reductions in Energy Intensity of the US Economy), energy intensity of the US economy declined about 1.7 percent per year, down from a decline rate of 2.7 percent per year. This historical observation can be seen in two ways. (1) energy intensity declines slowed down when the oil price dropped precipitously, but (2) energy intensity was still declining 1.7 percent per year during a time of relatively low oil prices. In addition, energy intensity has continued to decline by about 1.7 percent per year during the recent decade of relatively high oil price. That trend, the result of many forces operating together, and not simply the result of oil prices, appears to be quite stable. Thus the historical record suggests it is not unreasonable to expect continuation of energy intensity reduction of about 1.7 percent per year, even in the face of the recent precipitous drop in crude oil prices.

Other considerations are consistent with the expectation of a continued reduction in US economy-wide energy intensity.

Several factors imply continued decreases in energy intensity for transportation. Automobile makers are likely to meet the CAFE standards for new light-duty cars and trucks, although meeting the standards will impose some difficulties. But future US presidents could alter those standards, making them either more or less stringent. So presidential politics can create uncertainty for future automotive fuel economy.

---

11. Remember the old adage, cited above: "It's difficult to make predictions, especially about the future." Also, remember that anyone who knows what the price of oil will be in the future will never have to work for a living, because that person could reel in so much money from the oil futures markets. I have to work for a living!

Electric vehicles remain a tiny market share, but their sales can be expected to grow. California's zero-emission-vehicle mandate and low carbon fuel standard will accelerate market adoption in that state. Drivers find that electric vehicles are fun to drive, with exceptional performance. Maintenance costs for electric vehicles can be expected to be lower than costs for internal combustion engines. The electricity cost per mile for electric vehicles is substantially smaller than the gasoline cost per mile for internal combustion vehicles. And battery costs are decreasing significantly, reducing the cost premium for electric vehicles and leading to greater vehicle ranges between charges.

Electric vehicles use less energy per mile than do automobiles with internal combustion engines, so growth of electric vehicles will contribute to decreases in energy intensity of the economy.

In addition, as shown in Figure 2.5 (Vehicle Miles Traveled [trillions]: All US Roads), automobile travel is not growing as rapidly as it has historically. The combination of less growth in vehicle-miles-traveled and less fuel used per mile of driving suggests that total energy use for automobiles is likely to continue declining in the decades ahead.

New models of commercial aircraft have been introduced and existing models are being sold with more efficient engines. New Airbus and Boeing airplane families provide substantial fuel efficiency gains over models they are designed to replace. So air travel can be expected to continue to become more energy efficient. And it is in the economic interest of airlines to continue finding ways of filling all of their seats, so we can expect energy-efficiency improvements associated with increased capacity factors.

In commercial and residential lighting, we can expect increasing market penetration of LEDs. Currently LEDs have only a small market share of the installed base of lights. Estimates prepared for the Department of Energy are that LEDs account for only about 2.4 percent of the installed base of A-type lights (the typical screw-in light bulb), 5.8 percent of the installed base of directional lights, and 2.8 percent of all indoor lighting.[12] Much room is left for increased market penetration. In addition, conventional lighting is continuing to become more efficient as a result of regulatory standards.

Corporations continue to employ energy-efficiency professionals and sustainability officers. The financial benefits of reduced energy use typically exceed the cost of these employees. Corporate environmental transparency projects, such as the Carbon Disclosure Project, add to the incentive on corporations to reduce their energy footprint. Nongovernmental organizations such as NRDC and EDF continue to work with companies to find energy-efficiency opportunities. Thus I expect energy efficiency to continue improving in the corporate sector.

Even with no new federal governmental initiatives, state and local governments will continue to be active. Utility-funded energy-efficiency programs can be expected to continue, regulated by the states. More than half the states have adopted decoupling mechanisms for either electric or natural gas utilities. Building codes require progressively stricter standards for energy efficiency in new buildings. For example, California's Title 24 building standards continue to ratchet down the energy use that would be expected in newly constructed buildings. The New York State Energy Research and Development Authority, NYSERDA, has a wide variety of programs to promote energy efficiency. The six regional organizations founded to aggressively promote energy efficiency can be expected to continue operating. Several states— Arizona, California, Connecticut, Oregon, and Washington—set efficiency standards on appliances. Some of these standards, for example,

---

12. Mary Yamada and Kelsey Stober, "Adoption of Light-Emitting Diodes in Common Lighting Applications," Navigant, US Department of Energy, July 2015, http://energy.gov/sites /prod/files/2015/07/f24/led-adoption-report_2015.pdf.

those in California, have impacts well beyond the individual state, becoming de facto standards for the rest of the United States.

In short, energy-efficiency history and current ongoing activities suggest that going forward, the United States can continue to enjoy the many benefits of growing energy efficiency as long as it maintains the complex set of conditions that have enabled energy efficiency for over four decades. With careful nurturing of private- and public-sector energy efficiency, with appropriate pricing and policies, the trend of increasing energy efficiency can be accelerated, with further beneficial impacts on the environment, national security, and the economy.

I expect future United States energy-efficiency enhancements to continue the 40-year post-energy-crisis history, so energy efficiency will continue to shape the US energy system, bringing environmental benefits of economy-wide decarbonization, security benefits of net energy exports, and economic benefits of reduced costs.

Appendix A
Conversion Efficiency in Electricity Generation

Production in each sector of the economy employs electricity,[1] which in turn requires primary energy to generate. The measures of energy use by sector in this study include the amount of energy required to generate the electricity used by sector as well as the amount used directly (e.g., natural gas to heat a home).

If there were large increases in the efficiency of primary energy conversion to electricity, those increases would reduce the calculated amount of primary energy for each sector. But there has been only a relatively small increase in electricity generation conversion efficiency. Therefore, this possibility has not been a large contributor to the energy savings presented in this study.

This can be seen by examining the average heat rate for US electricity generation.[2] These data, published by the Energy Information Administration,[3] are shown in Figure A.1. Figure A.1 shows that the heat rate for electricity generation from fossil fuels, wood, and renewables (in black) and from nuclear (in red) declined about 10% from 1973 through 2014, with the largest decrease coming after 2000. This change in the heat rate contributed only a small fraction of the reductions in primary energy use, and that fraction occurred primarily after 2000. Therefore, almost all energy-intensity changes shown in the various graphs are based almost entirely on end use efficiency within those consuming sectors.

---

1. Electricity is often called "secondary energy." As defined by EIA, "Secondary energy sources are also referred to as energy carriers, because they move energy in a useable form from one place to another. There are two well-known energy carriers: Electricity, Hydrogen." http://www.eia.gov/energyexplained/index.cfm?page=secondary_home.
2. The heat rate is the millions of BTUs of primary energy used per megawatt hour generation of electricity.
3. Data source: http://www.eia.gov/totalenergy/data/monthly/pdf/sec13_6.pdf.

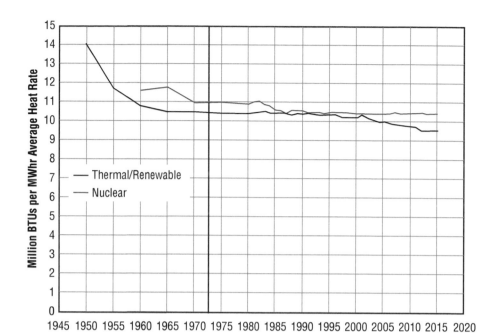

Figure A.1. Average Heat Rate for Electricity Generation

Source: Published by the Energy Information Administration

Appendix B
Calculation of Carbon Intensity of Energy Consumption

Carbon intensity of energy consumption is calculated by multiplying the carbon intensity of each primary energy source (measured in million metric tons per quadrillion BTUs) by its market share, and adding the products together. Data on carbon intensity of the primary energy sources and on market shares of energy consumption are from the US Energy Information Administration. Carbon dioxide emissions from nuclear, biomass, hydroelectric, wind, solar/PV, and geothermal are treated as zero for these calculations.

These data are in Table B.1 for two years, 1973 and 2014. Data for other years are calculated in the same manner.

Table B.1 also includes calculations of impact on intensity, which is the difference between the carbon intensity of the particular energy source and the carbon intensity of the total, multiplied by that particular source's fraction of all energy. For example, coal in 1973 averaged 93 million metric tons of carbon dioxide per quadrillion BTU of energy (MMT/Q), and the total of all primary energy had an average of 62.6 MMT/Q. The impact on intensity of coal is calculated as (93.0 MMT/Q—62.6 MMT/Q) × 17.1 percent, for an *increase* of 5.2 MMT/Q. Similarly, the impact on intensity of nuclear power is a *reduction* of 0.8 MMT/Q because the carbon intensity of nuclear power was less than the average. Thus the row "Impact on Intensity" is the impact on average carbon intensity of the particular energy source in the year, relative to the average carbon intensity of *that year*. So in 1973, the impact is in comparison to 62.6 MMT/Q, and in 2014 the impact is in comparison to 54.8 MMT/Q. For 2014, coal increased carbon intensity by 7.4 MMT/Q, and nuclear power decreased carbon intensity of energy consumption by 4.6 MMT/Q.

	COAL	NATURAL GAS	PETROLEUM	NUCLEAR	BIOMASS	HYDROELECTRIC	WIND	SOLAR/PV	GEO-THERMAL	TOTAL
1973										
Carbon Intensity	93.0	52.3	67.5	0.0	0.0	0.0	0.0	0.0	0.0	62.6
Fraction of Energy	17.1%	29.7%	46.0%	1.2%	2.0%	3.8%	0.0%	0.0%	0.0%	
Impact on Intensity	5.2	−3.0	2.3	−0.8	−1.3	−2.4	0.0	0.0	0.0	
2014										
Carbon Intensity	95.2	52.2	64.6	0.0	0.0	0.0	0.0	0.0	0.0	54.8
Fraction of Energy	18.3%	27.9%	35.4%	8.5%	4.8%	2.5%	1.8%	0.4%	0.2%	100.0%
Impact on Intensity	7.4	−0.7	3.4	−4.6	−2.7	−1.4	−1.0	−0.2	−0.1	
Impact on Intensity, 73 reference	6.0	−2.9	0.7	−5.3	−3.0	−1.6	−1.1	−0.3	−0.1	−7.7
Change in Impact	0.7	0.1	−1.5	−4.5	−1.8	0.8	−1.1	−0.3	−0.1	
Percent of Impact Changes	−10%	−2%	20%	59%	23%	−10%	14%	4%	2%	
Market Share Change	Increase	Decrease	Decrease	Increase	Increase	Decrease	Increase	Increase	Increase	

Table B.1. Calculation of Changes in Carbon Intensity of Energy Consumption[1]

1. Data from US Energy Information Administration, *Monthly Energy Review*, October 2015: Table 12.1 Carbon Dioxide Emissions from Energy Consumption by Source; Table 1.3 Primary Energy Consumption by Source.

In addition, there is a 2014 calculation based on comparison with the 1973 average intensity. This is based on the same calculation, but with the 1973 average intensity of the total energy consumption. Thus the row "Impact on Intensity, 73 reference" is the impact on average carbon intensity of the particular energy source in the year relative to the average carbon intensity *in 1973*. For example, the impact on intensity of coal is calculated as $(95.2 \text{ MMT}/Q - 62.6 \text{ MMT}/Q) \times 18.3$ percent, for an *increase* of 6.0 MMT/Q. Nuclear power *reduces* intensity of energy consumption by 4.6 MMT/Q.

The line "Change in Impact" is the difference between the 2014 "Impact on Intensity, 73 reference" (6.0 MMT/Q for coal) and the 1973 "Impact on Intensity" (5.2 MMT/Q for coal). This difference, 0.8 MMT/Q, tells the amount by which increase in the market share of coal and increases in the carbon intensity of coal have increased the carbon intensity of energy consumption between 1973 and 2014.

Similarly, for nuclear power, the change in impact is the difference between the 2014 "Impact on Intensity, 73 reference" (-5.3 MMT/Q for nuclear) and the 1973 "Impact on Intensity" (-0.8 MMT/Q for nuclear). This difference, -4.5 MMT/Q, shows that the increase in the market share of nuclear power has decreased the carbon intensity of energy consumption by 4.5 MMT/Q between 1973 and 2014.

The total impact of all of the changes together has reduced carbon intensity of energy consumption from 62.6 to 54.8 MMT/Q, a difference of 7.7 MMT/Q.[2]

The row, "Percent of Impact Change" shows the impacts of changes of each of the fuels as a percentage of the total, 7.7 MMT/Q. It shows that nuclear power has accounted for 59 percent of the total reduced carbon intensity of energy consumption, as a result of its increase in market share. Similarly, combustion of biomass has accounted for 23 percent

---

2. Rounding error accounts for the apparent inconsistency. The more precise numbers are 62.565 and 54.838, with a difference of 7.727 MMT/Q.

of the total reduced carbon intensity, also as a result of its increase in market share. Petroleum has accounted for a 20 percent reduction as a result of its *decrease* in market share. Wind, solar/PV, and geothermal have accounted for reductions of 14 percent, 4 percent, and 2 percent, respectively, as results of their market share increase. Working in the opposite direction, coal has accounted for −10 percent, that is, it has increased the average intensity by an amount equal to 10 percent of the total decrease in average intensity. Hydropower has also accounted for −10 percent, that is, it has increased the average intensity by an amount equal to 10 percent of the total decrease in average intensity, as a result of its *decrease* in market share.

About the Author

James L. Sweeney, known for his work on energy economics and policy, analyzes economic and policy issues, especially those involving energy systems and/or the environment. Particular interests include global climate change, energy efficiency, electricity markets, and energy market structure.

Sweeney is a professor of management science and engineering at Stanford University, director and founder of the Precourt Energy Efficiency Center, and a senior fellow of the Hoover Institution, the Stanford Institute for Economic Policy Research, the Precourt Institute for Energy, the US Association for Energy Economics, and the California Council on Science and Technology.

At Stanford, Sweeney was chairman, Department of Engineering-Economic Systems & Operations Research, 1996–98; chairman, Department of Engineering-Economic Systems, 1991–96; director, Center for Economic Policy Research, 1984–86; chairman, Institute for Energy Studies, 1981–85, and director, Energy Modeling Forum, 1978–84.

Sweeney's books include *California Electricity Crisis* (Hoover Institution Press, 2002), and *Handbook of Natural Resource and Energy Economics*, with A. V. Kneese (North Holland: Volumes I and II, 1993 and Volume III, 1995).

Sweeney earned a bachelor's degree in electrical engineering from the Massachusetts Institute of Technology in 1966 and a doctoral degree in engineering-economic systems from Stanford University in 1971.

HOOVER INSTITUTION

SHULTZ-STEPHENSON TASK FORCE ON

Energy Policy

The Hoover Institution's **Shultz-Stephenson Task Force on Energy Policy** addresses energy policy in the United States and its effects on our domestic and international political priorities, particularly our national security.

As a result of volatile and rising energy prices and increasing global concern about climate change, two related and compelling issues— threats to national security and adverse effects of energy usage on global climate—have emerged as key adjuncts to America's energy policy; the task force will explore these subjects in detail. The task force's goals are to gather comprehensive information on current scientific and technological developments, survey the contingent policy actions, and offer a range of prescriptive policies to address our varied energy challenges. The task force will focus on public policy at all levels, from individual to global. It will then recommend policy initiatives, large and small, that can be undertaken to the advantage of both private enterprises and governments acting individually and in concert.

Index

Page numbers in italics indicate tables and figures.